Home Workshop Guns for Defense and Resistance

Home Workshop Guns for Defense and Resistance

Volume III

The .22 Machine Pistol

Bill Holmes

PALADIN PRESS • BOULDER, CO

Also by Bill Holmes:

.50-Caliber Rifle Construction Manual:
   With Easy-to-Follow Full-Scale Drawings
Home Workshop .50-Caliber Sniper Rifle (video)
Home Workshop Guns for Defense and Resistance, Vol. I:
   The Submachine Gun
Home Workshop Guns for Defense and Resistance, Vol. II:
   The Handgun
Home Workshop Guns for Defense and Resistance, Vol. IV:
   The 9mm Machine Pistol
Home Workshop Guns for Defense and Resistance, Vol. V:
   The AR-15/M16

Home Workshop Guns for Defense and Resistance, Vol. III:
The .22 Machine Pistol
by Bill Holmes

Copyright © 1995 by Bill Holmes

ISBN 13: 978- 0-87364-823-3
Printed in the United States of America

Published by Paladin Press, a division of
Paladin Enterprises, Inc.
Gunbarrel Tech Center
7077 Winchester Circle
Boulder, Colorado 80301 USA
+1.303.443.7250

Direct inquiries and/or orders to the above address.

Visit our Web site at www.paladin-press.com

C ontents

WARNING

In 1994, the United States Congress, at the instigation of President Bill Clinton, passed a so-called "Crime Bill." This bill, while doing little, if anything, to prevent crime, contains a provision which prohibits further manufacture of any and all versions of the firearm described in this book. Govern yourself accordingly.

DISCLAIMER

Technical data presented here on the manufacture, adjustment, and use of firearms or ammunition inevitably reflects the author's individual experiences and beliefs with particular firearms, equipment, materials, and components that the reader cannot duplicate exactly. Before constructing or assembling firearms, gun parts, or accessories, care should be taken that no local, state, or federal laws are being violated.

The information in this book should be used for guidance only and approached with great caution. Neither the author, publisher, or distributors of this book assume any responsibility for the use or misuse of information contained in this book. It is *for academic study only.*

Introduction

Some 12 years ago I designed and built a compact, lightweight 9mm machine pistol which I called the MP82. This was done primarily to determine the feasibility of such a gun since, at the time, the military was soliciting proposals for a new submachine gun. The gun worked well enough, but I redesigned and refined it still further, resulting in another version which I called the MP83. I didn't get the military contract, although I sold several of the guns to government agencies at later dates. The contract was awarded to Heckler & Koch for a version of its MP5.

Even so, almost everyone who saw my prototype gun liked it and wanted one. So I managed to sell all I could make for the next several years.

In early 1983 I decided a .22-caliber gun based on the overall design of the 9mm gun was desirable. I wanted the gun in an open-bolt version, primarily because it required less parts and labor than a closed-bolt gun, but also because I had hopes of offering a full-automatic version to class-three dealers as well as military and police units.

After a small amount of redesign and refinement, the gun worked just the way I had hoped it would. This, in spite of advice (unsolicited) from a number of pseudoexperts who said that such a gun could never work dependably since cartridge rims, or heads, would hang up under the fixed firing pin and interfere with feeding.

I got around this problem simply by installing the firing pin at the lower edge of the bolt face instead of at the top. In this fashion, the firing pin actually pushed the cartridge ahead of it out of the magazine and into the chamber. It had simply never occurred to any of my critics that such a thing was possible.

This gun, which I called the MP22, was the most popular item I've ever offered for sale. I had a waiting list almost from the start. I built the gun in both full- and semiautomatic versions.

Eventually I sold a bunch of full-automatic silencer-equipped guns to a government agency. They had the idea that these weapons would be ideal for taking out sentries. Although I delivered the guns as agreed, I never got paid. Subsequent complaints and general hell raising on my part resulted in my guns being declared illegal and got me put out of business. But that's another story and dealt with in another book.

While the gun described in this book is basically similar to the original, I have made several changes. None of these are detrimental to performance. In fact, most could be considered improvements over the original

1

The Holmes MP22A1 .22 machine pistol.

The open-bolt assembly shown just above the gun is interchangable with the closed bolt.

Right side of the MP22A1.

design. There is only one threaded union on this gun, as compared to three on the original. The frame, or lower receiver, is folded and welded up from sheet metal; the original was machined from a solid block, which was time consuming, at best. Properly finished, this assembly looks as good, or better, than its predecessor.

The trigger mechanism incorporated into this gun can be considered an improvement, mainly because it allows the use of both an open- and closed-bolt configuration without changing any trigger parts. Switching bolts is all that is required. The two-stage trigger used in the full-automatic version is accomplished simply by adding an extra notch on the trigger bar and an extra spring.

By eliminating the vent holes in the barrel shroud, it is possible to incorporate a silencer, or suppressor, without any additional length.

Although it will be obvious to most readers, it should be pointed out that the gun is composed entirely of wood and steel. There are no plastic or pot metal parts. Even though a number of manufacturers and writers have attempted to brainwash consumers into believing that these materials are superior to wood and steel, most users soon find out otherwise.

It should be noted that the only version of this gun that would be considered legal by federal agencies is the closed-bolt version with a ventilated barrel shroud. The feds will maintain that the open-bolt version is too easily converted to full automatic. They will also say that an unvented barrel shroud version is almost a silenced version already. And if they should catch you with an open-bolt version with a two-stage trigger, you will be in big trouble. As a matter of fact, when the feds find out that this gun can be converted simply by switching bolts, you can be assured that they will try to declare any and all versions illegal.

So as long as commercial firearms are available, I would advise you not to attempt to build any version of this gun. When the government actually tries to take away all firearms, as it seems inclined to do, then may be the time to build your gun.

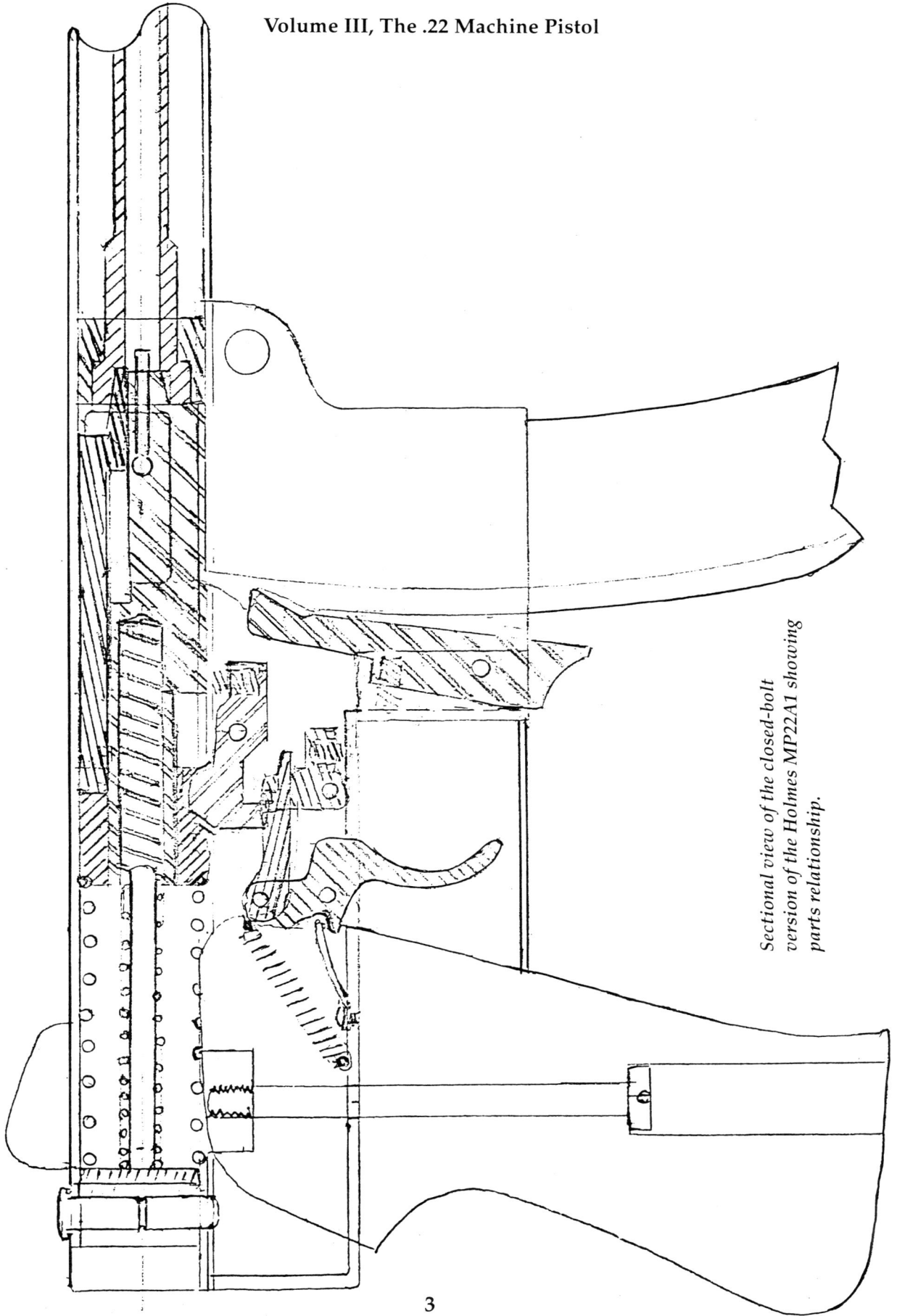

Sectional view of the closed-bolt version of the Holmes MP22A1 showing parts relationship.

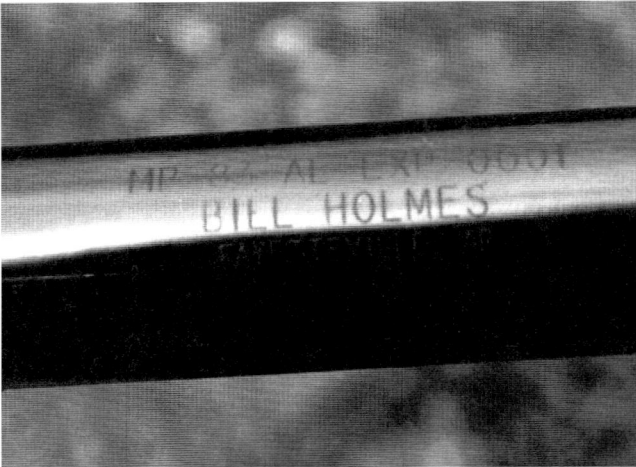

To comply with federal regulations at the time this gun was made, the gun had to have the maker's name and address, plus the model and serial number, inscribed on it.

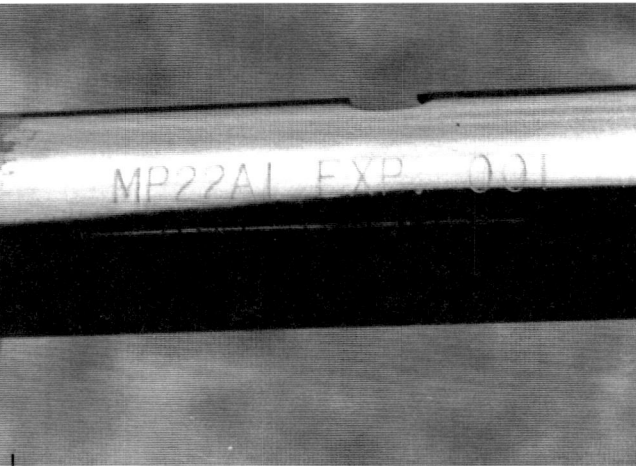

The gun portrayed in this book had serial number 001.

With a blued finish on the upper receiver and small parts, a case-colored lower receiver, and fancy English walnut grip, the gun attracts attention wherever it's displayed.

As matters stand today, I must state that this book is for academic study and information purposes only. Likewise, I must disclaim any responsibility for legal problems you may encounter if you fail to heed my warning. Since I have no control over your workmanship or the quality of materials you may use, I must also disclaim any responsibility whatsoever for any accidents, injuries, or safety hazards you may encounter.

1 Tools and Equipment

In several of my previous books, I have described in detail the equipment I've owned and used in my shop. Just to set the record straight, I no longer own most of this equipment. Our government has put me out of business twice in the last five years and cost me almost everything I had. Rather than losing even more money, I sold my shop and equipment and simply quit. But again, that's another story in another book.

I have since set up another small shop at my home with just enough equipment to build experimental and prototype pieces. I have no production capacity. I have no Federal Firearms License. If any federal agents want to come on my property, they will be required to obtain a search warrant before doing so. I intend to keep it this way.

Several weeks ago I bought a new 12 x 36 inch geared head lathe. This is a Chinese-made lathe with a cam-lock spindle, far superior to the belt-driven threaded-spindle lathes that cost almost as much. This one has a 1 5/8 inch hole through the spindle, which is the main reason I chose it over several others. While it is smaller than my previous machines, it will do the same work even though it takes more time.

About this same time I acquired a little Taiwan-made milling machine. It has a vertical head as well as a supported arbor for horizontal milling. This is not one of the little table-mounted heavy-duty drill presses with a milling table like we see advertised in the discount machinery catalogs but rather is a floor-mounted, knee-equipped machine. It is limited only by its short 30-inch table.

Against my better judgement, I bought one of the little 4 x 6 inch metal-cutting band saws that the discount houses offer for sale. This works fairly well as long as the blade is sharp. As soon as it gets the least bit dull, the blade jumps off. Still, it is easier than a hand hacksaw. But if I had it to do over, I'd buy a bigger one.

I also have a heli-arc welder as well as an oxy-acetylene rig. These items, plus various hand tools, make up my shop as it exists today.

The gun we are discussing here will require approximately two to four hours of mill work as well as a like amount of lathe work. Welding will require another 30 minutes. If you don't already own the equipment required, you should consider hiring somebody to do it. It shouldn't really cost that much.

If care is taken, most of the slots and openings can be cut with a small hand grinder such as a Dremel tool using one of the thin "cut-off" wheels. This would cut down on the milling machine time.

While this lathe is only a 12 x 36 inch machine, it has a 1 1/2 inch hole through the spindle, making it adequate for gun work. This lathe is a geared-head machine with a cam lock spindle. It costs almost $3,000.

This little milling machine is a combination vertical and horizontal mill. It also has a universal table as well as six geared speeds. This is a handy machine for making gun parts.

A vise with an 8-inch jaw opening can be useful for many purposes, including forming sheet metal. The vise shown here was purchased in 1940 by my father from a second-hand store for $7. It is far sturdier than those of current manufacture.

A small bandsaw such as this is only slightly better than a hand hack saw, bur as long as the blade is sharp, it saves a lot of hand labor. It can be used in both vertical and horizontal positions.

One item that I have never said much about before requires mentioning—a good heavy-duty vise. This should be an all-steel vise with at least 6-inch jaws. Heavy sheet metal (as used for the lower receiver of this gun) can be formed using such a vise, a heavy hammer, and suitable blocks. I will discuss this further in the appropriate chapters.

If possible, read my book *Home Workshop Guns for Defense and Resistance: Volume I, The Submachine Gun.* It describes several makeshift tools that are handy to have and don't cost much.

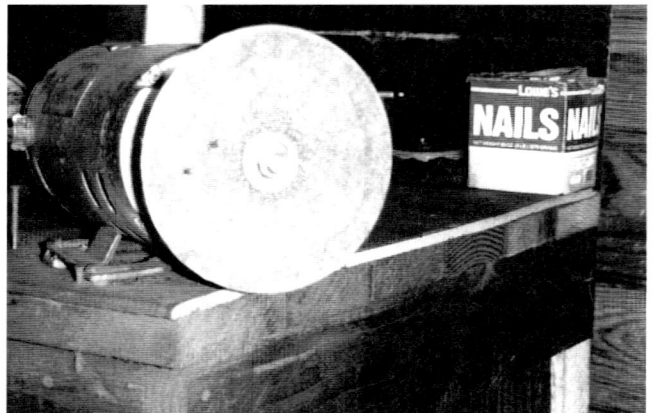

A 7-inch sanding disk mounted on a 1 horsepower motor can be used to shape both metal and wood.

Files in numerous shapes and sizes are essential.

This little hand grinder will cut all sorts of slots and openings. It is well worth the price.

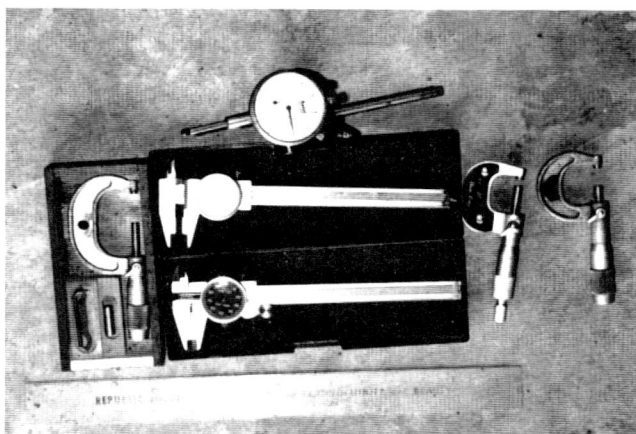

A few measuring tools are essential. Shown here are a Dial Indicator, 1- and 2-inch micrometers, Vernier calipers, and 6- and 12-inch rulers.

2 Materials

A suitable magazine for this gun is available (at the time of this writing) from Gun Parts Corporation, West Hurley, New York. The company calls these "Universal" magazines. They are used in Thompson .22-caliber guns as well as a number of others.

You can also obtain a barrel blank from this source. These will usually come slightly over 24 inches long, which means you can get four barrels from the one blank. Many times, gun shops have junk rifles that contain suitable barrel material.

You will likely be forced to purchase the upper receiver tubing from a metal supply house. Sometimes these places have short lengths available, but you may have to buy an entire stick just to get a piece 1-foot long. This can be expensive.

A sheet metal shop usually will have pieces of scrap material big enough to cut your lower receiver parts from. They will sometimes cut them to shape and bend them for a small fee.

Most of the other material needed can be found in a salvage yard. Automobile axles will furnish material for the bolt and other round parts. A leaf spring, as used on trucks and older cars, will furnish material for the flat parts. These will require annealing before they can be worked. I will discuss that further on.

Suitable springs are available from hardware stores and automobile supply houses. The recoil spring used in the original gun was intended for an M1 Carbine. These are available from military parts suppliers and can sometimes be found at gun shows.

A cabinet shop or furniture factory can usually supply hardwood for the grip.

3 Barrel

Since a barrel bushing—which must be machined to closely fit the breech end of the barrel, with a slot cut lengthwise to accept a locating pin—will be welded at a precise location inside the upper receiver, it makes sense to fabricate the barrel first and machine the bushing to match before welding it in place.

The barrel shown in the drawings has an overall length of 6.375 inches. Yours can be made longer or shorter by lengthening or shortening the lower receiver a corresponding amount. If only one gun is to be built, as it was in my case, the length can be as long as you care to make it. If several guns are contemplated, I suggest that the length be shortened to slightly less than 6 inches. This will allow four barrels to be cut from a 24 inch blank. At present, such barrel blanks are available from several makers, usually with an outside diameter of .875 inch.

It is sometimes possible to obtain a discarded rifle barrel from which a suitable length of barrel material can be salvaged. In certain instances, these may be smaller than the largest diameter of .700 inch shown in the drawing. This diameter can be reduced by as much as .075 inch if required, but a shoulder larger than the .500 inch diameter of the chamber area must be maintained.

Likewise, the diameter of the area over the chamber should be reduced very little, if any.

With the blank cut to length and the ends turned square, it is placed between centers in the lathe and turned to the desired contour. Many would-be machinists no longer believe it necessary to turn such a part between centers. They are content to catch one end in the lathe chuck, put a center in the other end, and go to work. This would be fine if the outer diameter were concentric with the bore. Many, or most, times, however, it is not. This results in the bore being off center, however slightly.

If it is desirable to turn it in this fashion, a short section at the centered, or tailstock, end should be turned slightly. The blank is then reversed and a section turned on the centered end. The blank is then reversed once more and turned to size since it is now as nearly concentric as possible using this method.

A portion of the barrel 1/2 to 3/4 inch long just behind the muzzle is threaded 3/8 inch x 24 threads per inch (TPI) to accept a corresponding nut, which, when drawn tight, holds the barrel in place.

A special cutting tool must be ground to form the approach cone in the breech end of the

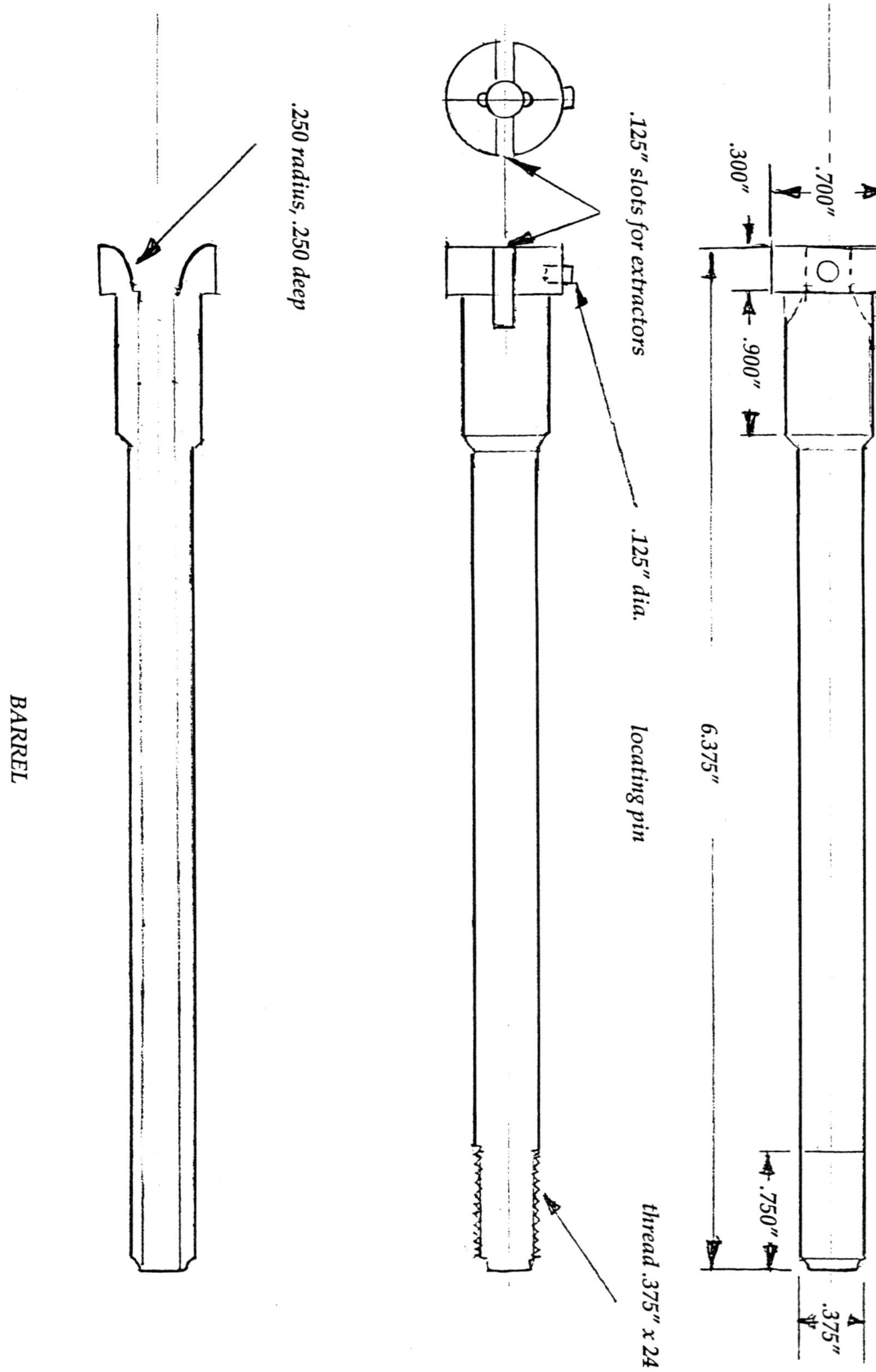

.250 radius, .250 deep

.125" slots for extractors

.125" dia. locating pin

thread .375" x 24

.300"

.700"

.900"

6.375"

.750"

.375"

BARREL

barrel. This tool is ground to a concave cutting edge with a .250 inch radius, or a .500 inch diameter portion of a circle. This tool is then set at a 30 degree angle and fed straight in. This forms a convex approach cone that is higher in the middle than at either end and will guide the bullet past the rim counterbore without shaving a sliver of lead off the bullet as it is fed into the chamber, as some other designs are noted for. The approach cone should be .250 inch deep, measured from the end of the barrel.

A hole for the locating, or indexing, pin is drilled just forward of the chamber end using a number 31 drill. A center drill should be used to start the hole and all other drilled holes incorporated in this design. The center drill will start straight without trying to wobble or crawl off, as is common with twist drills, especially after they become slightly dull. The pin hole should be drilled approximately .150 inch deep. A slight taper is ground at the extreme tip of a short length of .125 inch drill rod. This tapered end is started into the hole and pressed in to full depth using vise jaws or a clamp. Vise grips can be used for this if care is taken to prevent marring the barrel. This locating pin can only extend some .080 inch above the barrel surface, assuming the .700 inch barrel diameter is used.

Regardless of which bolt or firing pin is used, a clearance slot, or groove, must be cut at the bottom of the approach cone to clear the flange on the lower side of the bolt nose. If the closed-bolt version is used, a similar groove must be cut at the top to provide clearance for the firing pin. These cuts can't be located exactly until the bolt and upper receiver are completed. With the barrel in place, the feeding flange is coated with some sort of marking compound (lipstick will do). The bolt is then inserted in the receiver with the cocking handle in place and slammed forward a couple of times. This will mark the exact position. The cut is made by clamping the barrel in a vertical position with the chamber end up and cut with a 3/32 inch end mill. The firing pin relief cut is located and made in the exact same manner.

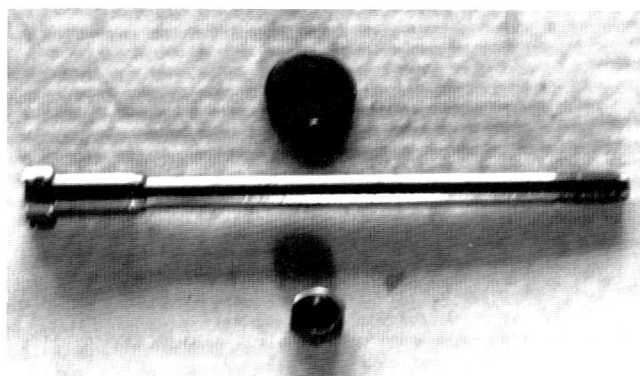

Barrel with muzzle cap and nut.

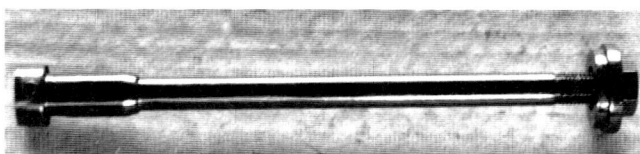

Barrel assembly with related parts in place.

Clearance cuts for the extractors are made in a like manner, except after the locations are determined and with the chamber cut to finished depth, the barrel is secured in the milling machine vise in a level, horizontal position, with the enlarged chamber end extending past the vise jaws. A 7/8 inch diameter by 1/8 inch thick keyway cutter is mounted in the collet and the extractor clearance cuts made from each side. This is something of a "cut and try" proposition. The cuts must provide clearance to allow both extractors to snap freely over the cartridge without interference. The clearance cuts must not extend into the chamber walls but must stop just even with the rim counterbore.

The chamber should be cut with .004 inch to .005 inch of clearance, or headspace, between the bolt face and the chambered cartridge. This can be determined by measuring the distance from the bolt face to the front of the enlarged portion of the bolt. This same measurement, plus the extra clearance, will be the distance from the breech end of the barrel to the head of the chambered cartridge.

4 Upper Receiver

The upper receiver is fabricated from a section of 4130 seamless tubing, 1 inch in outside diameter, and with a wall thickness of .065 inch. The slightly thinner .059 inch wall thickness can be used provided dimensions of the corresponding parts are altered accordingly. The specified length can also be changed to accommodate a shorter or longer barrel. Note that lengthwise measurements extend from the rear end forward. These will remain constant, regardless of barrel length.

With both ends squared and to the desired length, the ejection port is laid out and cut. This should be done before proceeding further so that the barrel bushing position can be observed and located. One half inch forward of the ejection port's front edge, four 1/4 inch holes must be drilled, equally spaced, around the circumference of the tube. The "equally spaced" part isn't terribly important since these won't show on the finished gun, but get them fairly close anyway, just for the practice if for no other reason.

The barrel bushing should be made before proceeding further. This should be turned to an outside diameter that requires a slight effort to push it into place inside the receiver. The slot must be cut long enough so the indexing pin in the barrel doesn't contact the end of it. The

barrel should slip in and out of the bushing with the indexing slot pointing toward top center. The bushing is pushed into position in the tubing and secured by welding through the four drilled holes. The welds are built up slightly above the surface and dressed back flush. Properly done, these welds will be invisible.

The breech and muzzle plugs can be turned to size at this time. They should be turned from material slightly over an inch in diameter due to the fact that round stock isn't always perfectly round and, when fitted into the ends of the tubing, might not make a concentric joint. The muzzle plug, or bushing, is simply bored and turned to size with a shoulder turned to slip inside the tubing. Note that the inside diameter may vary from the dimensions given, and the diameters of parts fitted into it may require revision. If care is taken, the seam will be seen as an almost imperceptible line when finished.

The breech plug, once turned to size, is installed in its respective position in the receiver and the .250 inch diameter hole for a retainer pin is drilled from top center entirely through both the top and bottom of the receiver and the breech plug, all at the same time. Upon completion of this, the plug is removed and another hole drilled through from one side perpendicular to and intersecting the existing

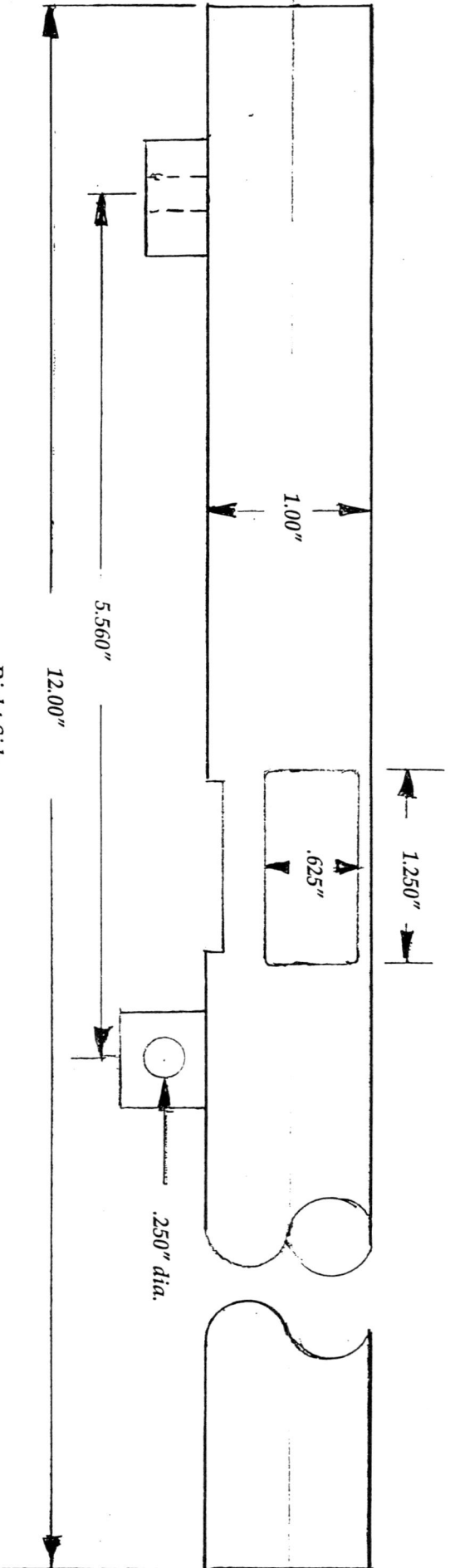

UPPER RECEIVER

Bottom

Drill and tap 1/4" x 28

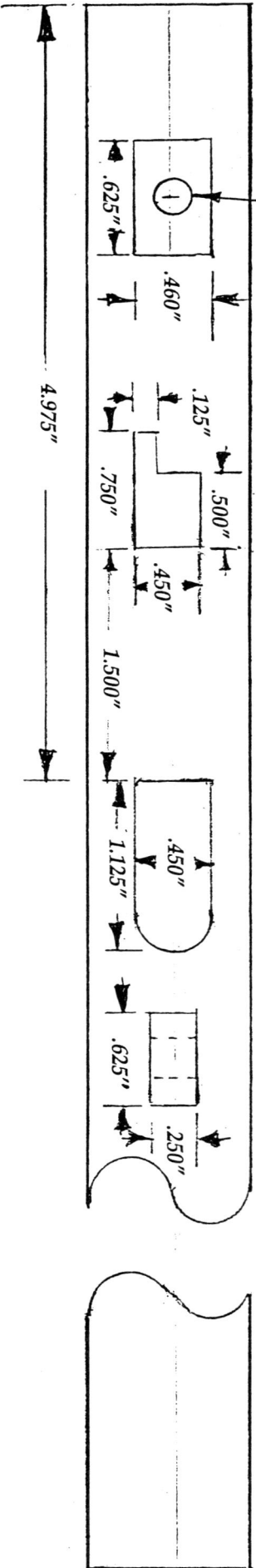

.625"

.460"

.125"

.750"

.500"

.450"

1.500"

4.975"

1.125"

.450"

.625"

.250"

Right Side

1.00"

5.560"

12.00"

1.250"

.625"

.250" dia.

.625"

3.000"

.250"

.250"

1.875"

.250"

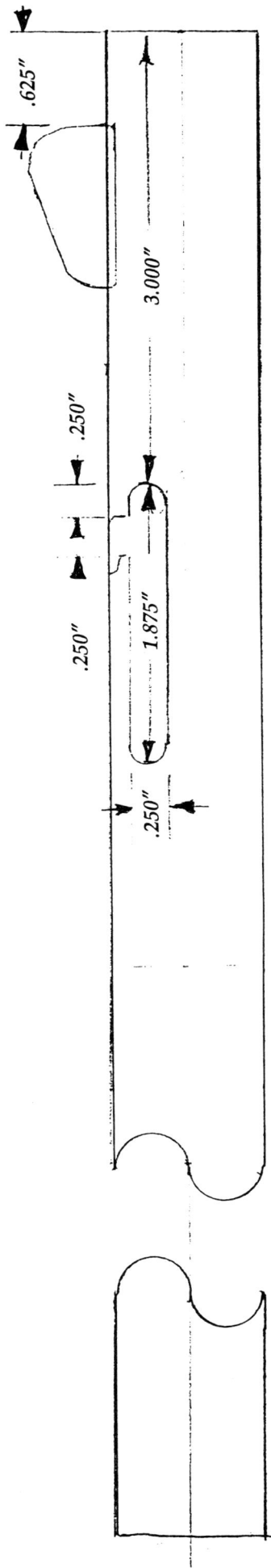

Drill four .250" holes around circumference. Weld barrel
bushing in place through holes.

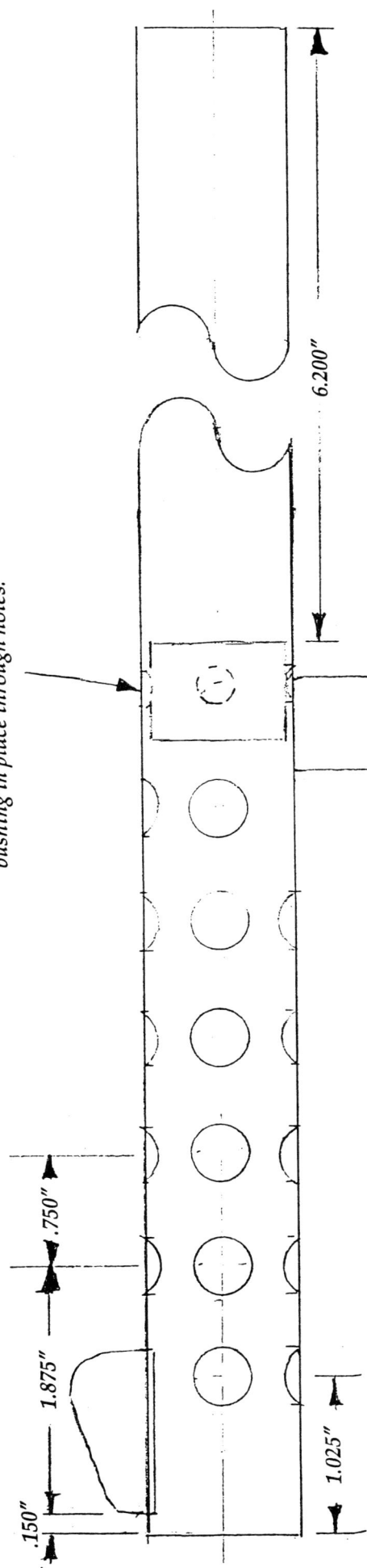

6.200"

.750"

1.875"

.150"

1.025"

UPPER RECEIVER, LEFT SIDE

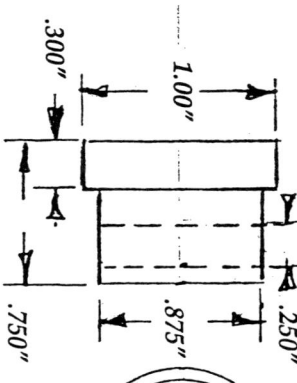

BREECH PLUG PIN

.300" 1.00" .500" 1.100" .312"

BREECH PLUG

.750" .875" .250" .250"

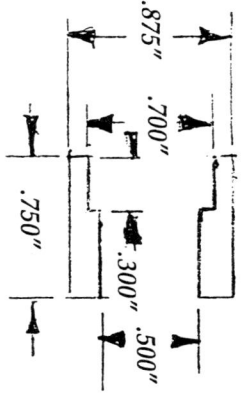

SPRING PLUNGER
8 x 32 x .375"

Slot .125 x .300"

BARREL BUSHING

.875" .700" .750" .300" .500"

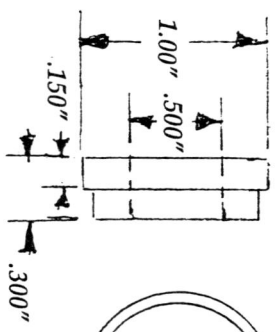

MUZZLE PLUG

1.00" .150" .500" .300"

RECOIL SPRING GUIDE

.860" 2.250" .125" .875" .187"

NUT

.875" .150"

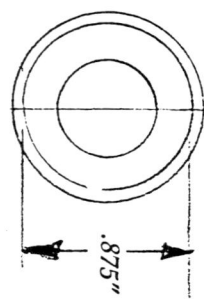

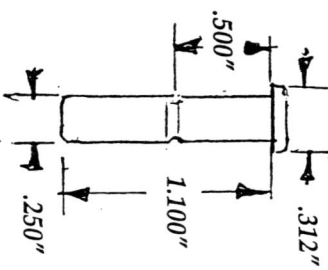

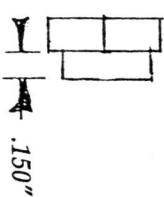

hole. This one is drilled with a number 27 drill and tapped to receive an 8 x 32 x 3/8 inch spring plunger. Such spring plungers are available from machine tool supply houses and sometimes from hardware stores. The spring-loaded plunger will engage a groove in the retaining pin and hold it in place. Remember to remove the spring plunger prior to bluing the plug.

The barrel retaining nut can be hand made or made the lazy way. The commercial 3/8 inch diameter by 24 thread nut I bought at a hardware store for this purpose measured just under 7/16 inch (.5625 inch) and was .325 inch thick. The hexagon nut was turned round part way, forming a round portion .150 inch long by .500 inch in diameter. The nut cost 10 cents and was machined in less than five minutes. Try to obtain a blue or black nut if possible, since chrome or cadmium plating requires too much work to prepare for bluing.

Two small blocks are welded in their respective positions along the bottom side. These will later be drilled and/or tapped to accept the bolt and screw that hold the upper and lower receivers together. The sides of these blocks that adjoin the tubing should be beveled and the groove built back flush by welding. If the 1/4 inch block is left slightly thicker and reduced to the correct width after welding, removing traces of the weld at the same time, a neater job will result. Don't drill the bolt holes at this time; wait until the lower receiver is built and drill them simultaneously, with both parts clamped together. This will ensure absolute alignment.

A slot for the bolt handle is cut in an approximate 10 o'clock position as viewed from the rear. The exact angle isn't terribly important since the cocking lever hole in the bolt will be used as a reference point to mark its location. The upright slot at the rear serves as a safety and bolt hold-open device. When the bolt handle is pulled to the rear and rotated upward into the slot, it moves forward slightly, holding it securely in place. This will work equally well on the closed-bolt gun, although it is a bit clumsy. However, this gun is not intended as a fast-draw weapon, so a slightly clumsy safety shouldn't matter much.

Four 1/4 inch holes are drilled around the circumference of the upper receiver intersecting the barrel bushing.

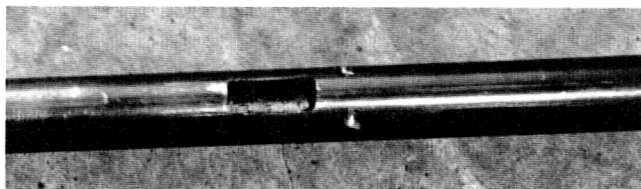

Holes are welded, fusing bushing to receiver body. If welds are built up above the surface and dressed back flush, no evidence of welding will be seen.

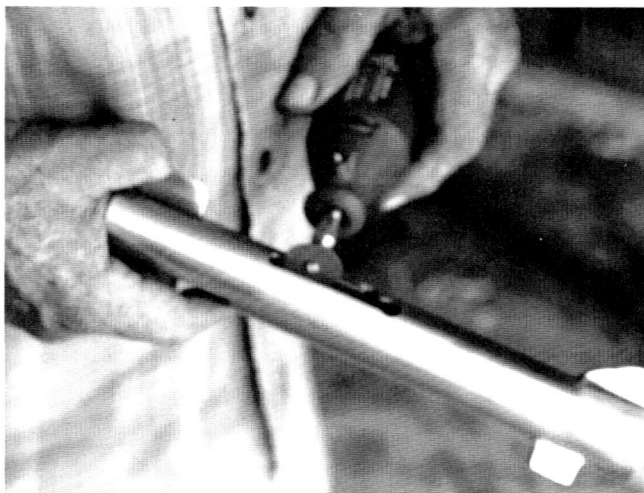

Openings are cut for the cocking lever, magazine opening, ejection port, and trigger assembly. This is done using end mills in a milling machine, but the same result (although slower) can be achieved with a hand grinder, as shown here.

Openings for the sear and disconnector can be cut at this time, but the magazine opening should be postponed until the lower receiver is finished and in place.

While they have nothing to do with function except to maybe keep the barrel cooler, four rows of ventilating holes should be drilled, equally spaced for looks, in the barrel shroud. This will hopefully prevent overzealous federal

Breech plug and retaining pin.

Breech plug with retaining pin inserted partially.

Blocks for mounting brackets are welded on lower side.

The hole for the breech plug retaining pin is drilled through both receiver and plug simultaneously.

The front screw hole is drilled simultaneously through the upper and lower receivers while they are clamped together.

Tapping for the screw is done in the same manner. Absolute alignment is assured using this method.

Vent holes in the barrel shroud are spaced accurately using the milling machine. If a ball cutter is used, smooth, round holes can be formed through both the top and bottom surfaces at the same time.

agents from maintaining that your gun is set up for a potential silencer.

It is very easy to incorporate a silencer into a gun of this design if the barrel shroud is left solid. All that is required is a series of ports drilled through the barrel and the area between the barrel and shroud packed with some sort of absorbent material. Steel wool or wire mesh scouring pads work satisfactorily for this. The feds are aware of this, so if you make up a gun that anyone else will ever see, this should be kept in mind.

5 Lower Receiver

Two identical sides are cut from 12 gauge sheet metal. This material has a nominal thickness of .104 inch, which can be expected to finish to approximately .100 inch after polishing. This material is too thick to cut with hand shears, so unless a commercial shear is available, probably the easiest way to cut it is with a fine-toothed hand hacksaw. I suggest that the two halves be cut slightly oversize to allow for misalignment or crooked saw lines.

Scribe a visible line at the points where the bends will be made. One of the side blanks is clamped between the vise jaws with the scribed line just visible above one jaw. Using a heavy hammer and a forging block, which consists simply of a heavy block of square or rectangular steel, the flange is bent over at a 90 degree angle. It should now be positioned flat against the top of the vise jaw. The other side is formed by the same method except that the flange is bent in the opposite direction. The flanges at the rear end are bent in the same fashion. Projection of the flanges should be measured. The magazines I have on hand measure .450 to .460 inch thick. The flanges should measure .230 inch from the inside. Measure your own magazines and adjust this measurement to accommodate them. The flanges, as formed, should be slightly wider than necessary. Clamp each side in the mill vise and mill off enough material to reduce them to the correct width.

A spacer is required to hold the two sides at the stated width while welding. This can be made by milling or grinding some .040 inch from one side of a 1/2 inch thick steel bar. One can be built up by using a piece of 1/4 inch thick material plus two pieces of 12 gauge sheet metal. This will give us a thickness of .458 inch. A strip of paper placed between these should make up the rest. The two sides are clamped in position with the spacer in between and the seams welded. As usual, the TIG or heli-arc process is preferred for this. Once done, the welds are ground or machined back flush with the parent metal and dressed smooth.

A front piece, or block, is made as shown in the drawing. Unfortunately, this part must be as thick as the sides are wide, so it is likely that a piece of 3/4 inch material must be reduced to the required thickness. The radius on the inside can be plunge-cut by mounting the part vertically and using a 1/2 inch end mill or mounting it horizontally and using a 1/2 inch ball cutter. The slot is cut with a 3/32 inch end mill. These cuts can be made with a round file for the radiused part and a hand hacksaw with two blades mounted if no milling machine is available. The part is positioned with the reduced-width section between the two sides, the width checked, and, if correct, welded in place. The magazine should now slide in and out with very little side play. The openings for the trigger and magazine latch should be cut before proceeding further.

The magazine latch housing serves a dual purpose both as a mounting point for the latch and to limit rearward travel of the magazine. The housing is formed by bending the sheet metal blank around a 1/4 inch thick form block, using the vise and hammer as before. When sized and shaped as shown, the part is placed inside the rear of the magazine opening and, with the magazine in place, pushed up against it until the magazine has no fore and aft play but will still slide in and out of the opening. It is then welded or silver soldered in place.

With all welds ground or otherwise dressed flush and smooth, the assembly is secured in the mill vise with the bottom side parallel to the vise bottom. The top side is first milled flat and then cut to fit the contour of the upper receiver using a 1 inch ball cutter. The overall height of 1.150 inch should be maintained as closely as possible simply to make the trigger parts locations come out right. A close-fitting slot to accept the front mounting bracket of the upper receiver is cut using a 1/4 inch end mill. The rear block should have been reduced in width from its original half inch when the welds were machined smooth to a width that will fit inside the lower.

Both parts should now fit together.

With both the upper and lower receivers clamped together, the hole for the front mounting screw is drilled, threaded, and bored for the screw head. With the screw in place, the rear bolt hole is located and drilled. Drilling the holes in both parts simultaneously assures that they match. The opening for the magazine is marked through the lower receiver and cut to size in the upper. This should match closely the dimensions shown in the drawing.

All pivot pin holes can now be located and drilled. Use a number 31 drill through both sides, then a 1/8 inch drill through only one side. The slightly smaller number 31 holes will secure the pins and keep them tight when they are forced in place. Matching pins can be cut from 1/8 inch drill rod slightly longer than the receiver width and rounded on each end.

The trigger guard, if you can call it that, is easily made simply by cutting a strip of 12 gauge sheet to the dimensions shown and welding the forward end in place. The rear end slips into a slot cut in the grip, which, when bolted in place, holds it securely.

The welded sheet metal receiver described here can be fabricated in far less time than one machined from a solid block. And for all intents and purposes, it is just as solid and sturdy. If the welding is well done, no one will ever know that it is not a machined assembly.

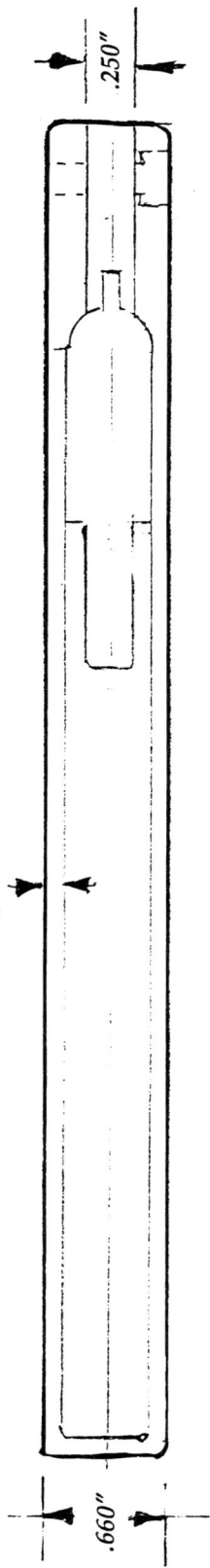

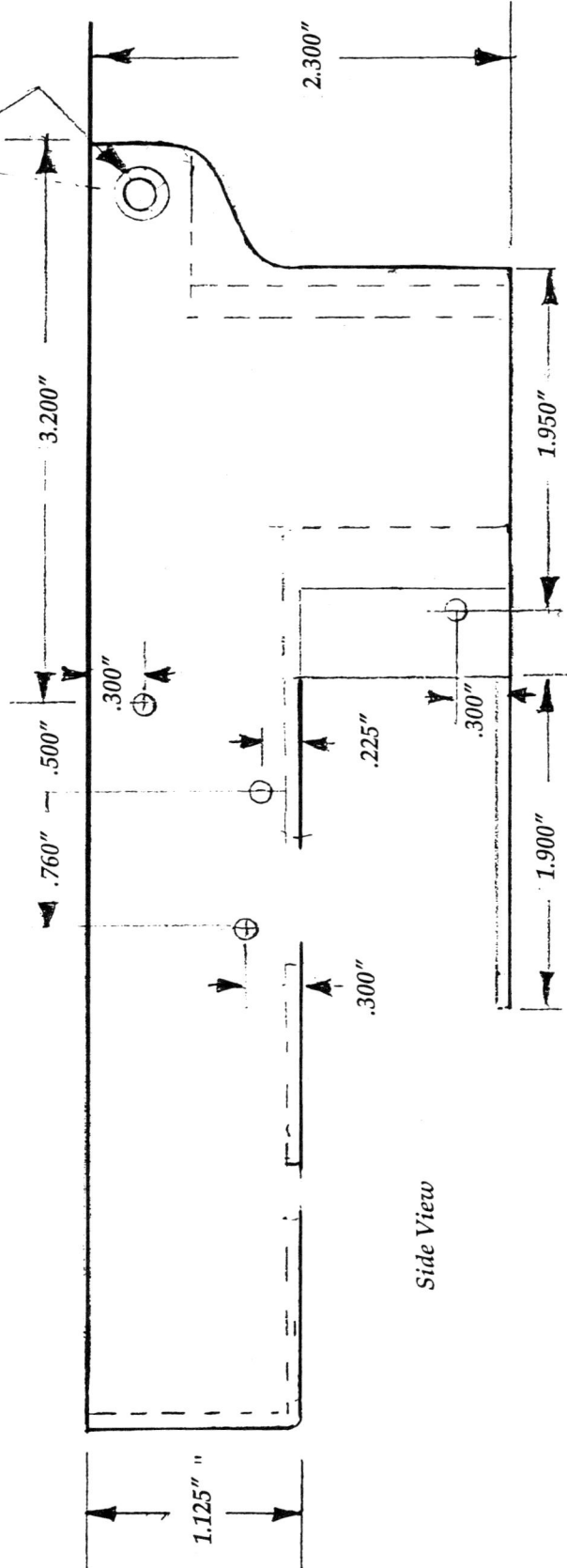

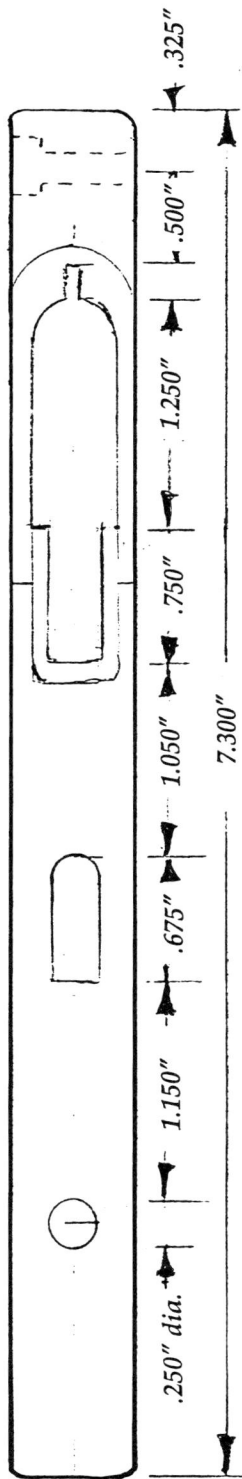

12 ga., .104"

.250"

Top View

.660"

10 x 32 screw

2.300"

3.200"

1.950"

.760"

.500"

.300"

.225"

.300"

1.900"

.300"

1.125"

Side View

.325"

.500"

1.250"

.750"

.750"

1.050"

.675"

7.300"

1.150"

.250" dia.

Bottom View

LOWER RECEIVER

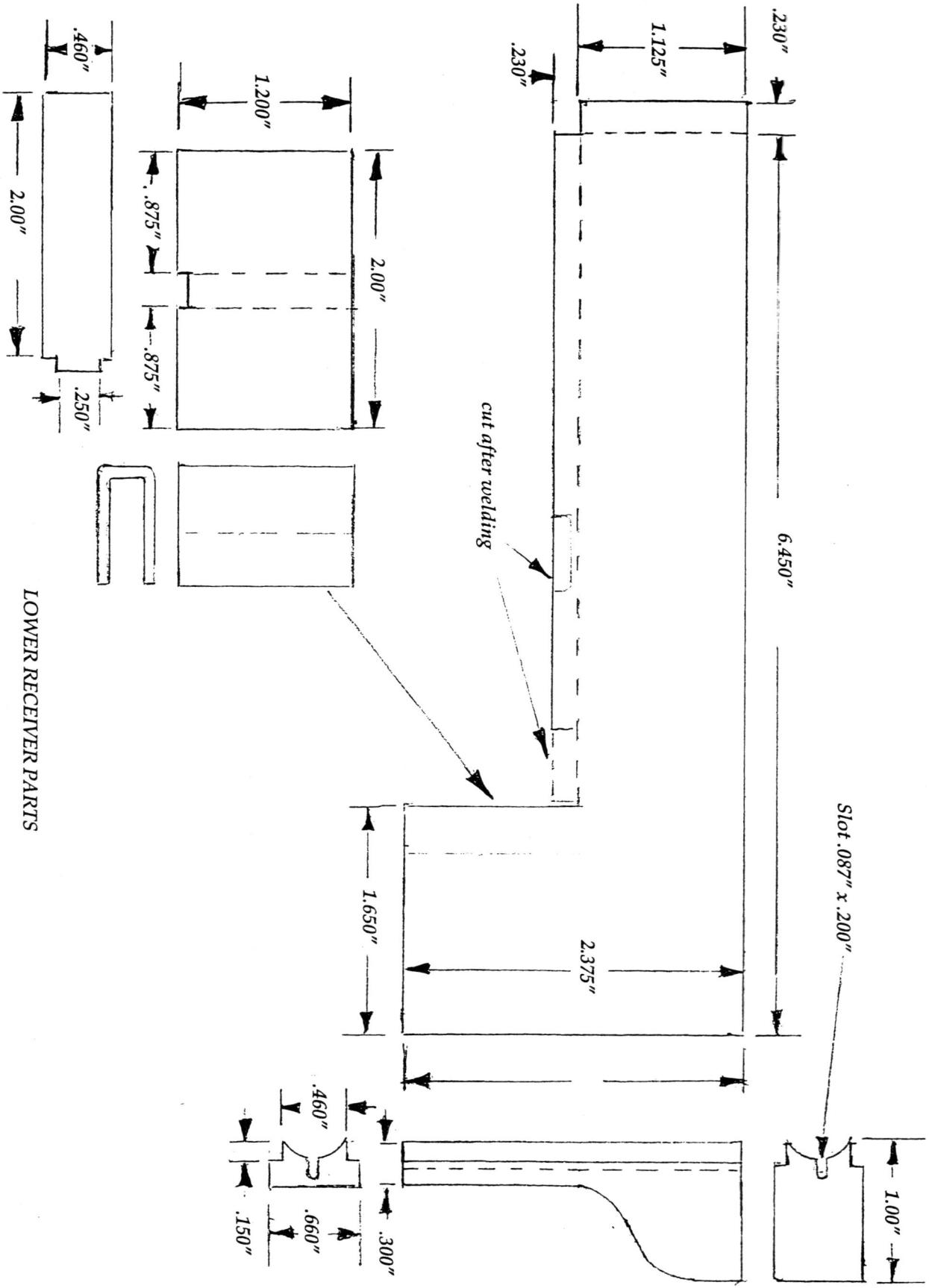

LOWER RECEIVER PARTS

.460"
2.00"
.250"

1.200"
.875"
.875"
2.00"

.230"
1.125"
.230"
6.450"

cut after welding

1.650"
2.375"

Slot .087" x .200"

.460"
.660"
.150"
.300"
1.00"

The pivot pin hole location can be located easily by placing the magazine latch in the assembled frame and inserting the magazine. With the latch pushed up against the magazine and held in place, the hole is drilled using the previously drilled hole in the frame as the locating point.

.250"

.125"

1.650"

.850"

.600"

Spring pocket
.204" dia. Use #6 drill.

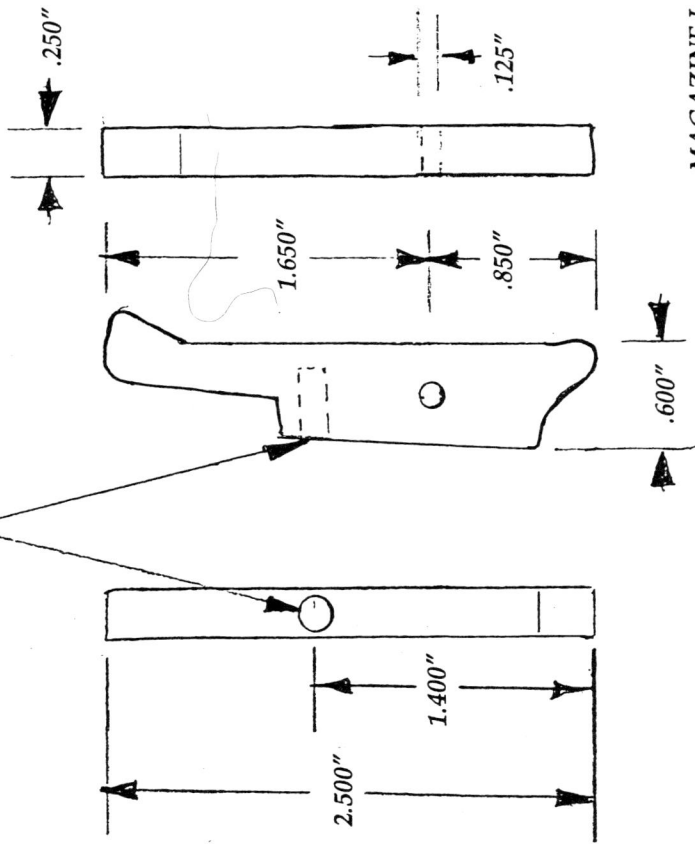

1.400"

2.500"

MAGAZINE LATCH

Lower receiver blanks are cut to approximate shape.

The blank, together with the form block, is clamped in the vise. Note that the form block is supported underneath to keep it in place.

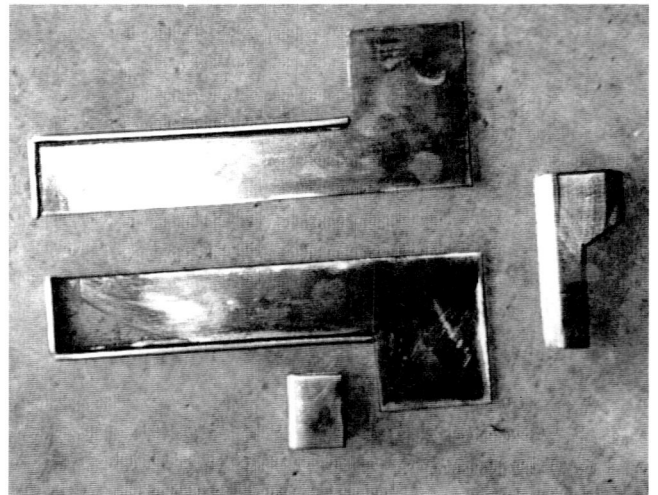

The flange, which will form the lower side, is folded 90 degrees with a heavy hammer.

An inside contour to match the magazine shape is cut in the front block, using a ball cutter for concave portion and a 1/8 inch end mill for slot.

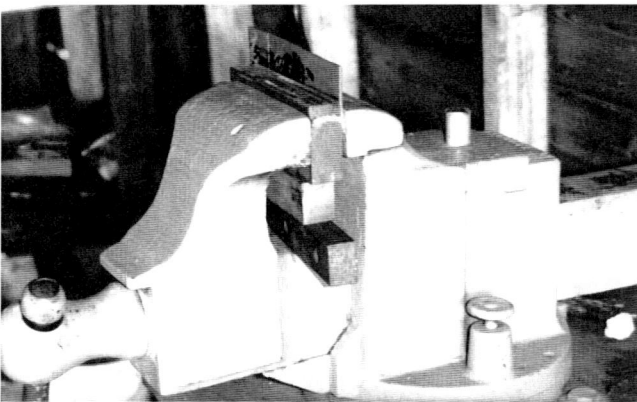

Component parts of lower receiver before assembly.

Clamped in place for welding. Note that a filler block is inserted to keep the assembly square and maintain dimensions.

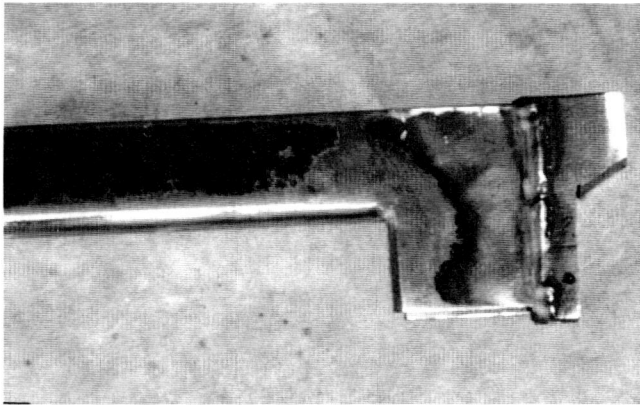

Welding complete.

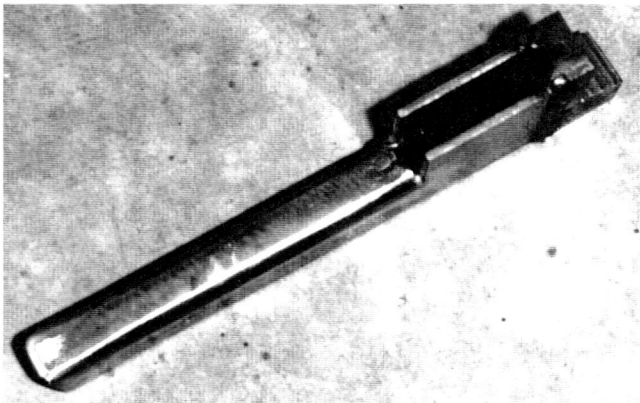

Lower seam is welded lengthwise.

Welds are ground flush with disk sander.

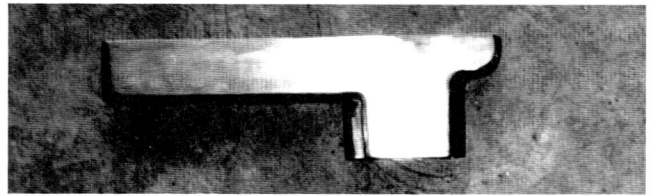

Lower receiver needs only final polishing.

Inside is cut to same radius as upper using 1 inch ball cutter.

6 Bolt

Although two different bolts are shown, one for a closed-bolt and the other for an open-bolt gun, both are similar as far as general dimensions go.

The open-bolt version is desirable, as far as I am concerned, only because it requires less parts and, therefore, is simpler and easier to make. Federal agents frown on these because, they say, they are easily converted to full automatic. While this may be true in some instances, most closed-bolt autoloaders can also be easily converted, usually with no more than a couple of hours work. In fact, the closed-bolt version of this gun could be made into a full automatic simply by adding one small lever to the trigger mechanism.

The open bolt, as previously stated, is far easier to make than the closed bolt. It should be made from a good grade of steel that will resist wear and battering. 4140 or 4340 works well for this. Car or truck axles contain suitable material. These are usually almost glass hard on the outer surface with a softer inner core. They will require annealing, or softening, before they can be machined satisfactorily. I have described a simple way of doing this in at least two other books, but at the risk of being considered repetitious, I'll describe it once more for those who haven't read the other books.

A large wood fire is built by gathering up old lumber, tree limbs, small logs, scrap wood, and whatever else is available and placing the material to be annealed on top of the pile. Leaf spring material for making flat parts should be included. The wood pile is then ignited and allowed to burn completely. The metal parts will usually fall into the ashes and coals, where they should be covered using a rake or shovel and allowed to cool slowly, overnight preferably.

If you own a bluing setup, metal to be annealed can be suspended a short distance above the burner, with the tank removed, and heated. This probably won't heat material as thick as an axle to the bright red required to completely anneal it, but it will soften it to a point where it can be machined readily.

The open-bolt version is made by turning an adequate length of material to a size that will slide freely inside the receiver. This should be .010 to .020 inch smaller than the inside diameter of the receiver.

A smaller bolt nose, which enters the barrel counterbore, must be formed at the forward end of the bolt. This should project .250 inch from the front face of the bolt body, have a diameter of .270 inch at the face, and angle 10 degrees toward the front. This may have a slightly odd appearance, but it results in the most foolproof

.22 rimfire feeding system I have experienced.

A slot to provide clearance over and around the magazine is cut with a 1/2 inch end mill, with two more narrow parallel slots cut to clear the magazine lips and ejector. A strip .150 inch wide is left in the center of the slot with .125 inch slots on either side. Note that the slot on the ejector slide must be deeper to provide clearance for the ejector.

Another small slot .094 inch wide must be cut along the middle (.150 inch wide) section and a combination firing pin and feeding flange silver soldered into it. This should be made from quality material which will not deform or batter easily. A short piece of 3/32 inch drill rod, ground flat on opposing sides to just fit the slot, is ideal for this. Note that this flange/firing pin combination extends below the bolt face and is flush with the center section. This is mandatory to assure that the bolt picks up cartridges reliably from the magazine. When finished, the firing pin should extend, or protrude, some .035 inch from the bolt face.

Extractor slots are cut on each side of the bolt using a 3/32 inch end mill. At the rear end of each of these, a .125 inch hole for a spring pocket is drilled to the same depth as the slot. Holes for the extractor pivot pins are drilled in their respective locations using a 3/32 inch drill. Since these holes originate on a steeply curved surface which will cause even a center drill to deflect to the lower side, these should be started with a 3/32 inch end mill, followed by a small center drill and then a standard twist drill.

Recoil spring diameter will dictate the size hole required to accept it. M1 Carbine springs measure from .255 to .260 inch in diameter. A 9/32 inch drill measuring .2812 inch will do for this, with up to .300 inch acceptable. This hole

should be 2 inches deep. Bevel the edge slightly to assist in guiding the spring.

The finished bolt is inserted in the receiver in its exact up and down position. The hole for the bolt handle is marked through the slot, the bolt removed, and the hole drilled. Enough clearance must be provided to assure that the handle does not contact the forward end of the bolt slot.

The closed bolt is made in the same fashion and to the same dimensions as the other at the forward end, except that no firing pin protrusion is allowed. The feeding flange must still be installed, but the forward end is flush with the bolt face. A section at the rear end is reduced to .500 inch diameter, guiding the striker, which encircles it. A 1/8 inch slot must be cut lengthwise in the top side of the striker, as shown in the drawing. This mates with the firing pin, which is silver-soldered in place. Another slot is cut on the bottom side, forming a shoulder to engage the sear. A third slot behind the magazine clearance cavity is required to clear the sear and a slot cut at the top, lengthwise, to receive the firing pin.

Better material than sheet metal must be used for the firing pin. A strip of the leaf spring material, milled to a 1/8 inch thickness, is desirable for this.

The extractors, likewise, should be made from the same material. These, however, are cut thinner, to a .0938, or 3/32, inch thickness. Note that only the extractor on the ejection port side is a true extractor. The one on the inside has an angled hook to permit the ejector to kick the cartridge rim out from under it. Its only purpose is to hold the case rim in its proper location against the bolt face and against the true extractor as it is withdrawn from the chamber.

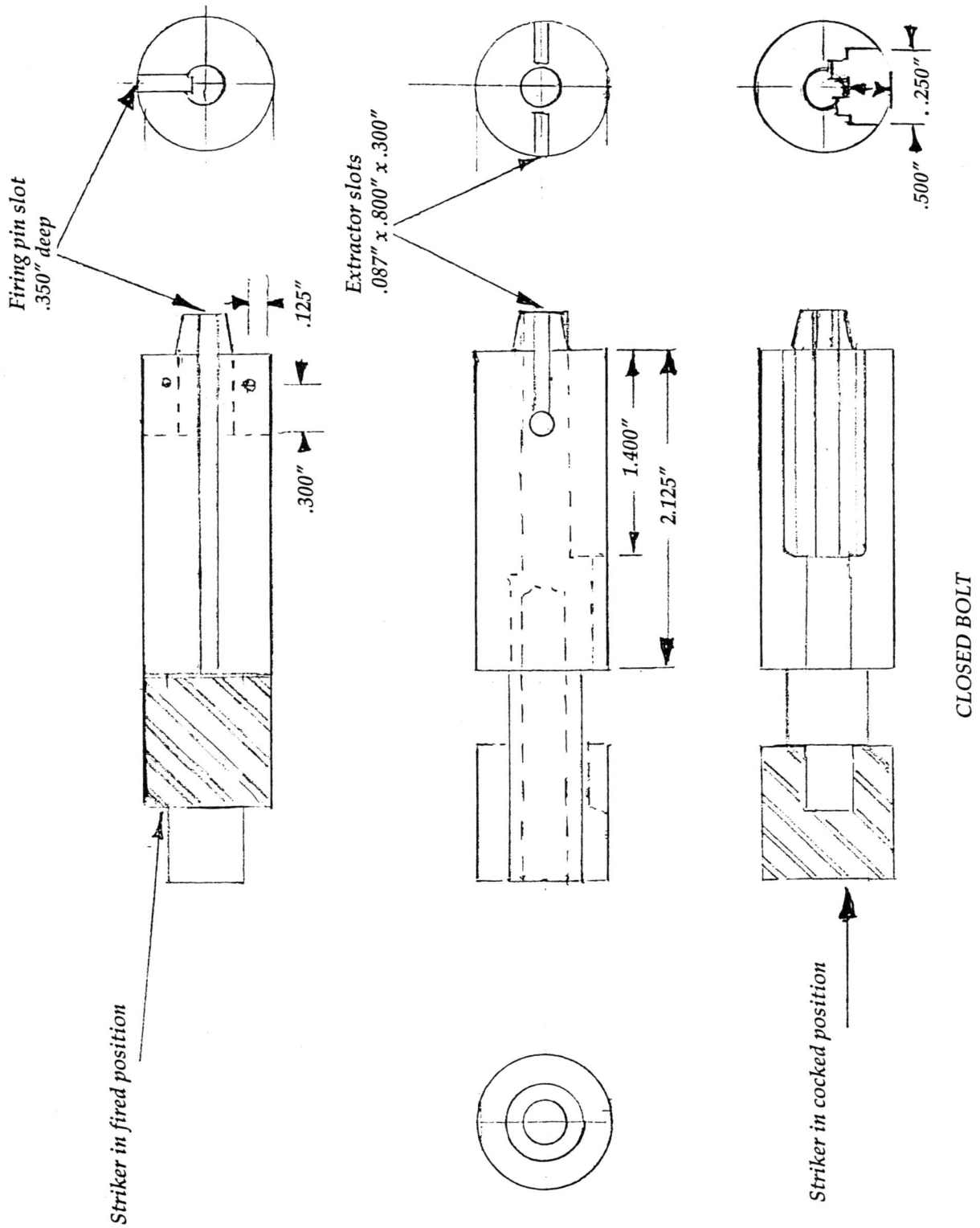

Firing pin slot
.350" deep

.125"

.300"

Striker in fired position

Extractor slots
.087" x .800" x .300"

1.400"

2.125"

.250"

.500"

CLOSED BOLT

Striker in cocked position

OPEN BOLT

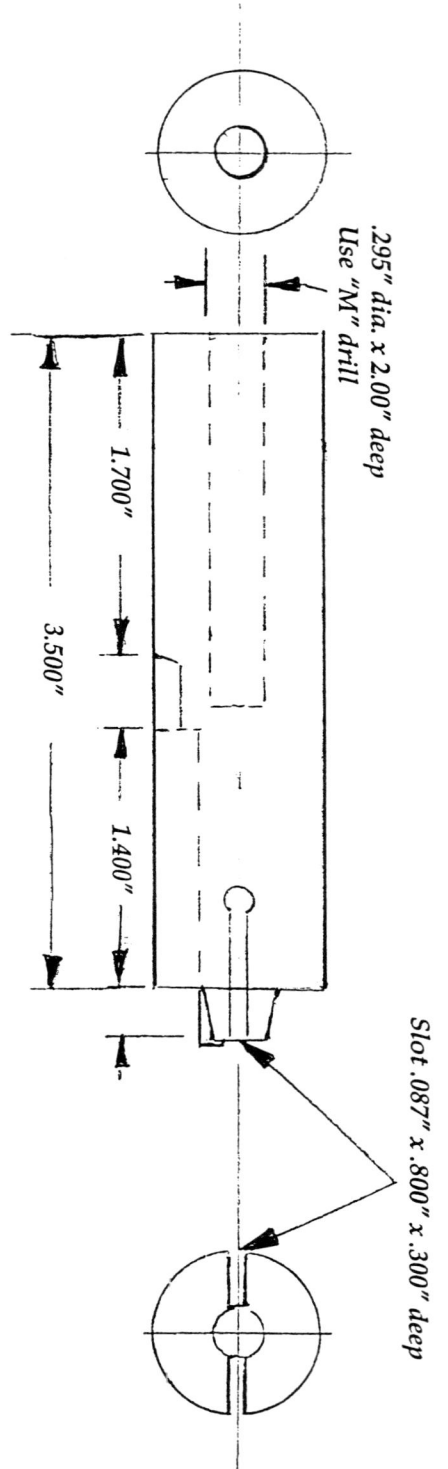

.295" dia. x 2.00" deep
Use "M" drill

1.700"

3.500"

1.400"

Slot .087" x .800" x .300" deep

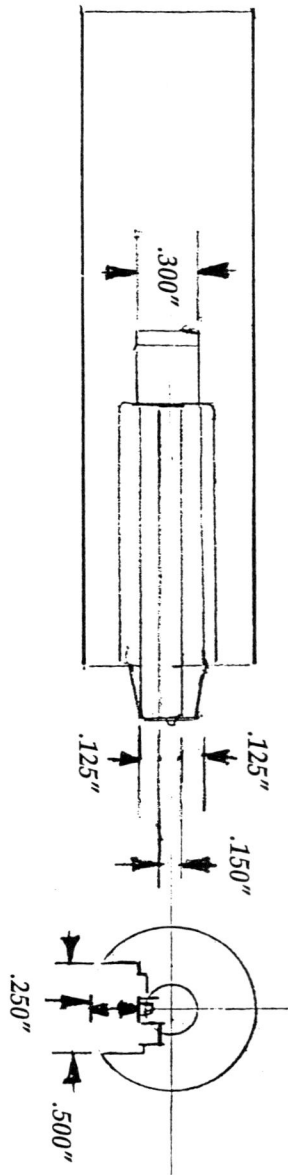

.300"

.125"

.125"

.150"

.250"

.500"

FIRING PIN

2.750"

3.00"

.150"

.600"

1.00"

1.50"

.065"

.250"

Slot .125" x .300"

.500"

.375"

.875"

STRIKER

.860"

BOLT BODY

3.500"

1.375"

.150"

EXTRACTORS

.125"

.800"

.250"

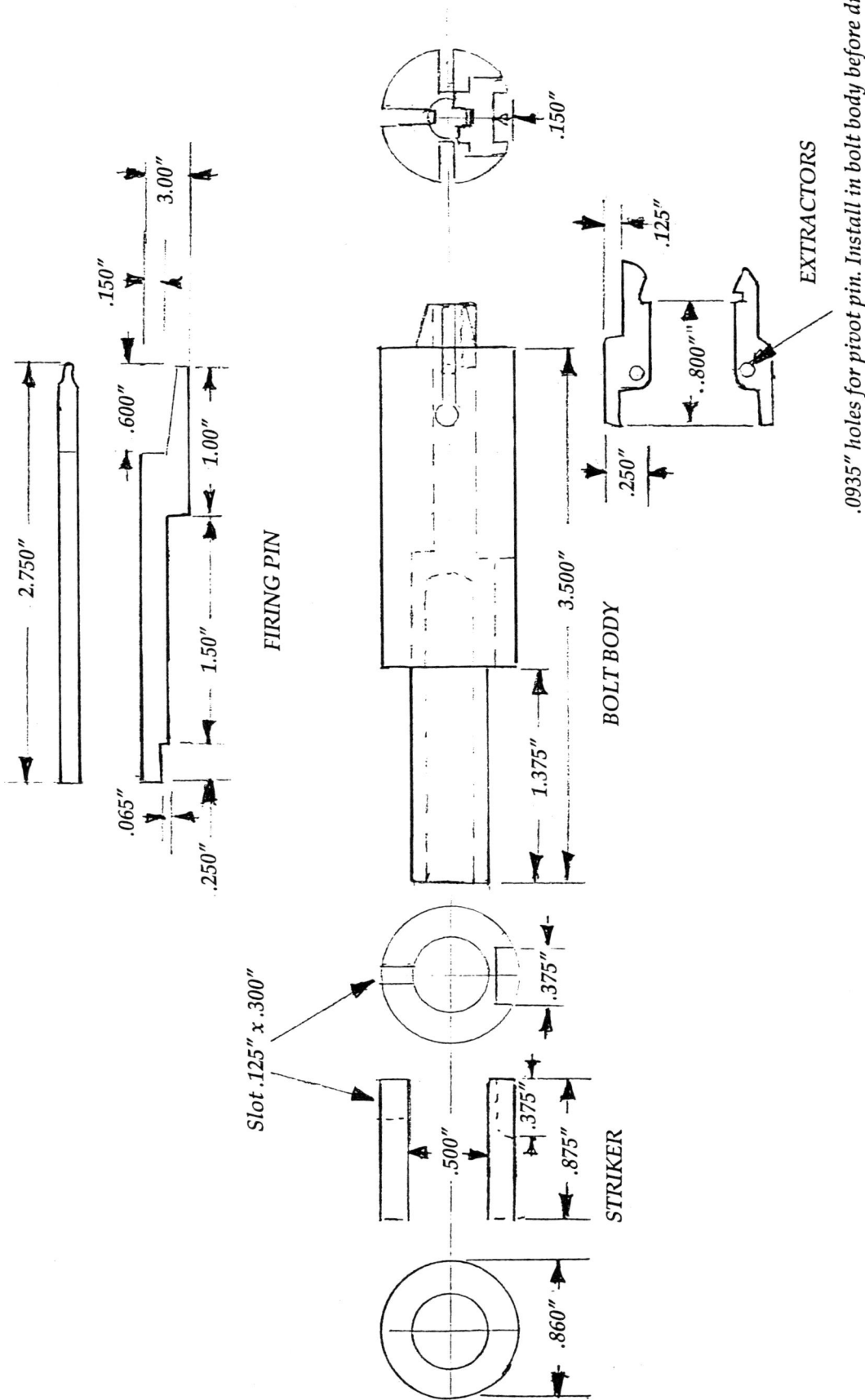

.0935" holes for pivot pin. Install in bolt body before drilling.

CLOSED-BOLT COMPONENT PARTS

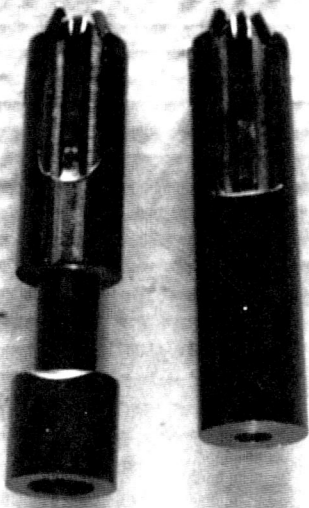

Interchangable bolts: closed bolt on left, open bolt on right.

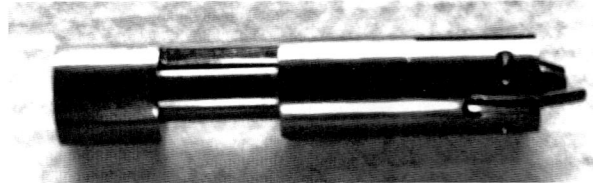

Closed bolt with springs and spring guide in place, ready for installation.

Disassembled closed bolt showing striker, extractors, and bolt body.

Closed-bolt assembly.

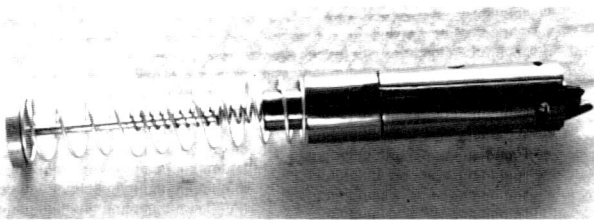

Open-bolt assembly.

Open-bolt assembly; fixed firing pin is on lower side of bolt face.

7 Trigger

The trigger assembly described here is composed of four parts plus pins and springs. While slightly unorthodox as compared to others in guns of this type, it is designed to function equally well in both open- and closed-bolt configurations. With only slight modification it will serve as a two-stage trigger for a full-automatic version whereby a short, light pull will fire single rounds, functioning as a semi-automatic, and a longer pull against a heavier, stiffer spring will cause the gun to continue to fire as long as the trigger is held back. Needless to say, this would translate into a highly illegal firearm. Don't say I didn't warn you.

Two of the parts, the sear and trigger block, must be made from material capable of being hardened. The trigger requires no heat treatment and can be made from any available steel. The trigger bar is cut from 12 gauge sheet metal.

It can be helpful, in some instances, to make cardboard patterns or templates of parts such as these. These are placed on the material to be used and drawn or scribed around, leaving an outline of the part to be cut. It can be helpful to drill the pivot pin hole first and lay out the shape of the part around it. This is far easier than trying to drill a precisely located hole in a finished or semifinished part.

The trigger is cut from 1/4 inch flat stock.

The trigger blade is cut first. It can either be sawed into a narrow strip and the curve bent into it, or the front face can be formed with a 1 inch diameter end mill. This is done first simply because you need enough material on the other end to hold on to while forming it. The front face can be crowned or rounded slightly using a half round file or, if available, a high-speed grinder of the Dremel tool variety. The upper leg can be formed by milling or sawing it to shape. The only critical dimension on this part is the distance between the pin holes, and even this need not be exact. The rest can be shaped to suit the maker. The projection at the rear is only required for the full-automatic version.

The trigger bar is cut from sheet metal and formed into an open-ended rectangle by bending it around a 1/4 inch form block. This part should be cut slightly oversize and finished after bending. The tab at the upper center with the small hole should be bent to the rear at a 30 degree angle. The coil pull spring connects to this and not only serves as a trigger return spring but exerts upward pressure on the forward part of the trigger bar. Exact dimensions cannot be given since this part must be hand-fitted during assembly.

The sear should be wide enough, from the pin hole forward, to prevent sidewise movement when mounted in place. The leaf spring material suggested may not be thick enough for this. If material of sufficient thickness is not available, a spacer can be soldered or cemented on one or both sides to increase the thickness. These should encircle the pin in the same fashion as a washer, which, when assembled, cannot work loose. Note that the spring pocket is all that keeps the spring in place during assembly, so it should be at least as deep as shown.

The trigger block, for lack of a better name, should also be almost as wide as the inside of the frame. This too can have a spacer or washer, but only on the left side as viewed from the rear. The small projection at the upper right hand corner is engaged by the notch on the trigger bar which pushes it out from under the sear when the trigger is pulled, causing the gun to fire. Here again, the overall shape is not terribly important.

While we are engaged in making these parts, we may as well go ahead and make up the magazine latch. This can be formed with the milling machine or sawed to shape from 1/4 inch thick flat stock. Here again, the overall shape is not that important. The pivot pin hole should not be drilled first as with the other parts but is drilled with the gun assembled and the magazine in place. The latch is pushed up firmly against the retaining notch in the magazine and the hole drilled. This will result in a near perfect fit without any cutting and trying, as is required when fitting some of the other parts.

With the parts in a semifinished condition, dummy pins should be cut long enough to go completely through the frame and protrude 1/2 inch or more from one side. Grind a slight taper on one end of these pins to allow them to enter the slightly smaller holes on the off side without undue resistance. The protruding ends should be rounded slightly. The trigger bar is installed on the trigger and pinned in place. Then, all the parts are installed in their respective positions but on the outside of the frame. It should be obvious what is required to fit these parts to a

point where they work the way they are intended.

The only part that should require much fitting is the trigger bar. With the trigger block positioned in place under the sear and the trigger in an approximation of its forward-most position, the notch at the front of the trigger bar must rise into position behind the projection on the trigger block without binding. Mark the notch location and cut it back to the mark. When it seems to work correctly, the parts should be mounted in their respective positions inside the frame and tried. Springs can be kept in place during assembly by filling the spring pockets with wheel bearing grease. Too much metal removed from the trigger bar notch will result in excessive trigger travel. Removing a small amount of metal from the face of the trigger block will allow it to move forward, reducing the gap between the notch and the trigger bar projection. Several tries may be required before you get this perfect. But when, and if, you do, a short, light trigger pull will result.

If the two-stage trigger is to be installed, another notch must be cut some .060 to .100 inch behind the existing notch and approximately the same distance above it. The .065 wire spring is bent to shape and installed so that it almost contacts the projection at the rear of the trigger. This is simply to cause enough resistance to distinguish between the first and second stages of the trigger. The harder, longer pull against this spring causes the second notch on the trigger bar to push and hold the trigger block out of engagement with the sear, allowing the bolt to reciprocate freely. This will not work with the closed bolt in place since the cartridge rim will hang up under the firing pin, interfering with feeding.

One more time: this trigger mechanism, in combination with the open bolt, will result in an illegal firearm. I have described it simply to illustrate the principle. I do not advocate that you build one. The closed-bolt gun should be adequate for your needs.

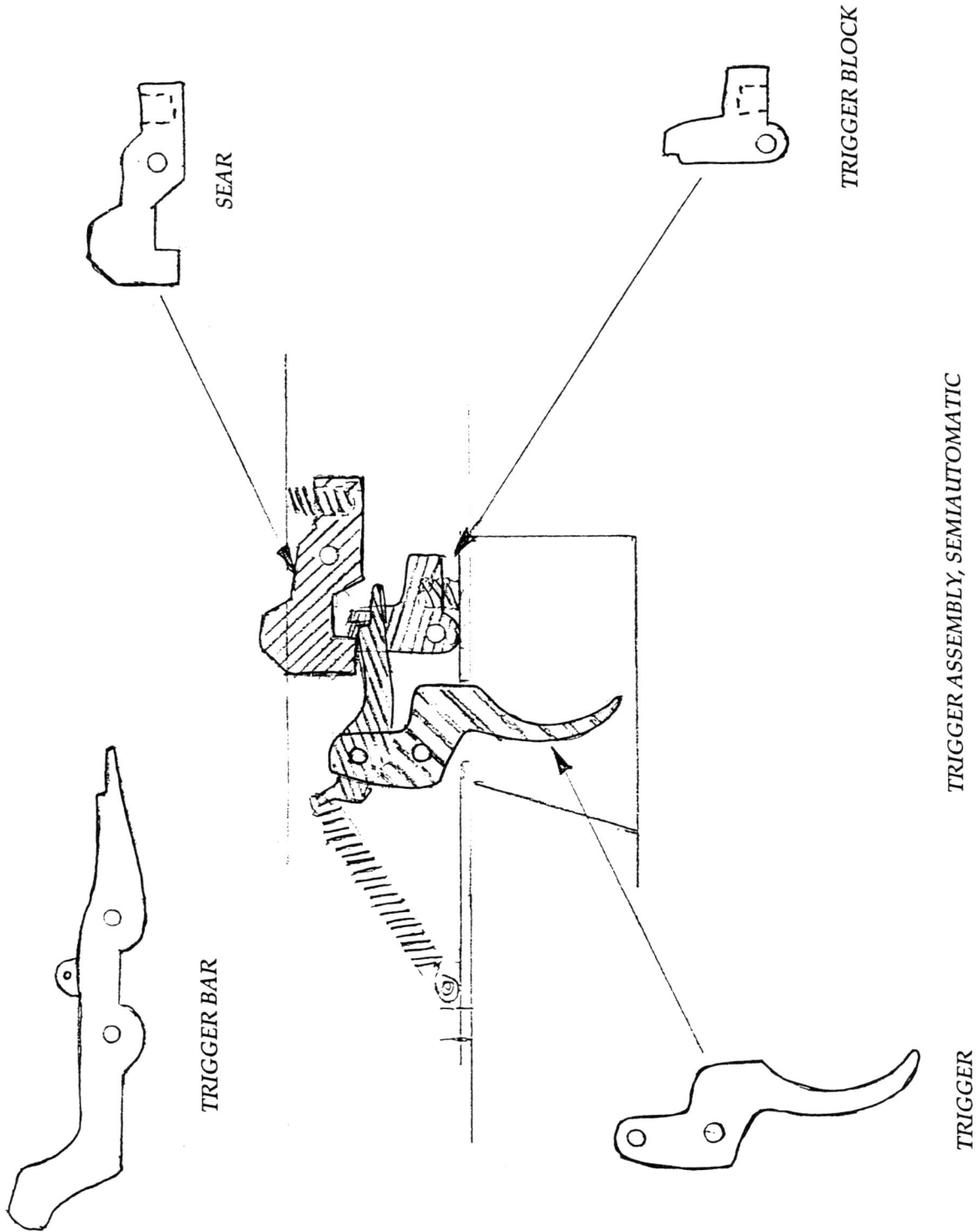

SEAR

TRIGGER BLOCK

TRIGGER BAR

TRIGGER ASSEMBLY, SEMIAUTOMATIC

TRIGGER

SEAR

TRIGGER BLOCK

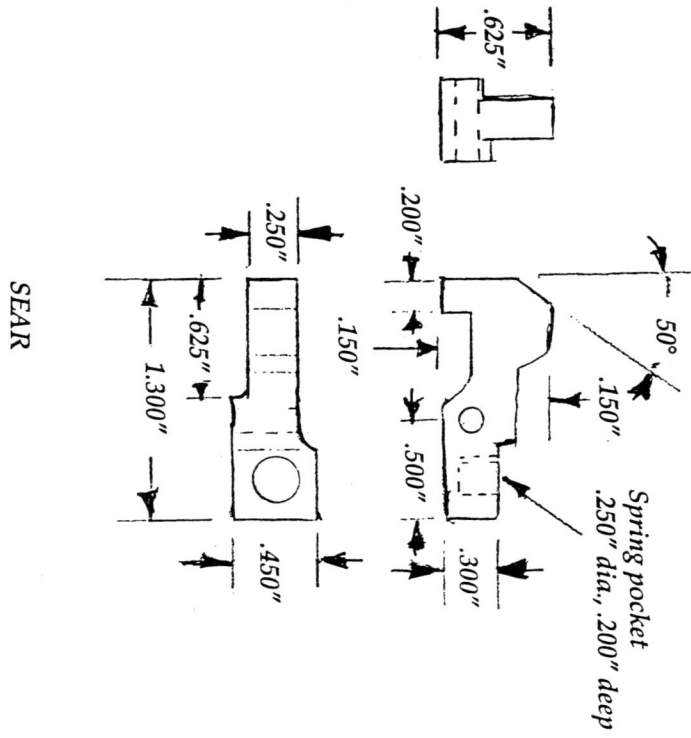

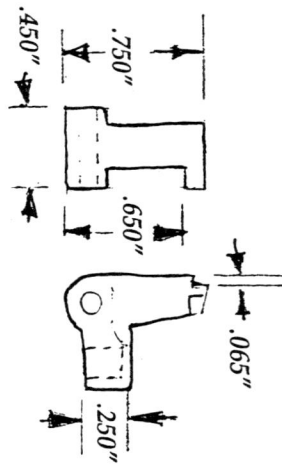

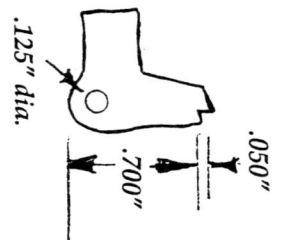

TRIGGER ASSEMBLY COMPONENTS

.625"

.250"
.200"
.150"
.625"
1.300"
.150"
.500"
.450"
.300"
50°
Spring pocket
.250" dia., .200" deep

.450"
.750"
.650"
.065"
.250"

.125" dia.
.700"
.050"

.065" dia.

.875"

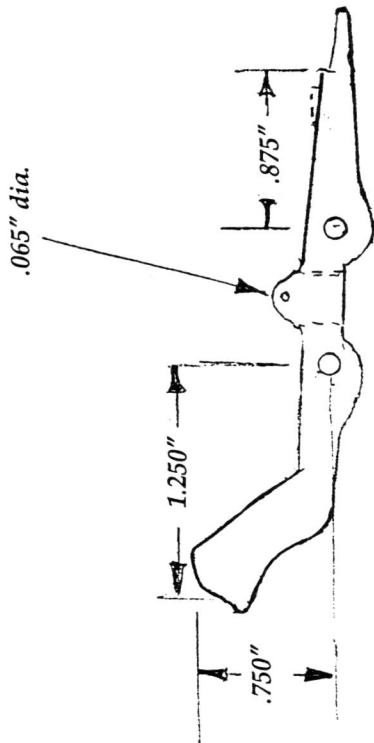

Fold to this shape.

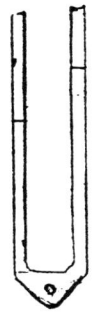

1.250"

.750"

TRIGGER BAR

Pull spring

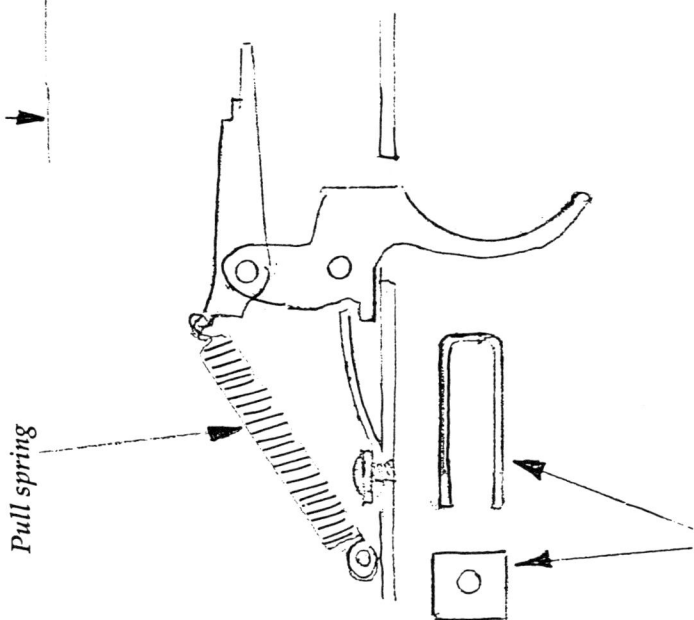

Spring bent from .065" music wire. Held in place by screw
and washer. Used only in full-auto version.

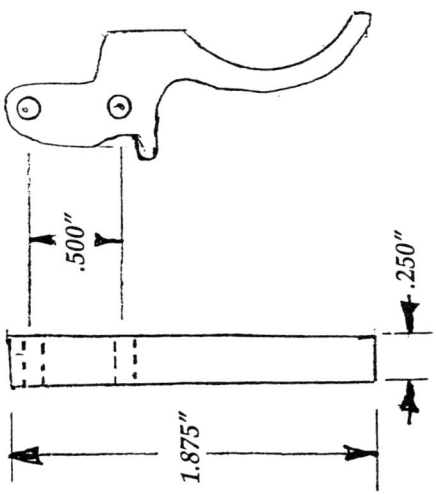

TRIGGER ASSEMBLY COMPONENTS

.500"

.250"

TRIGGER

1.875"

Component parts of trigger assembly.

Sear.

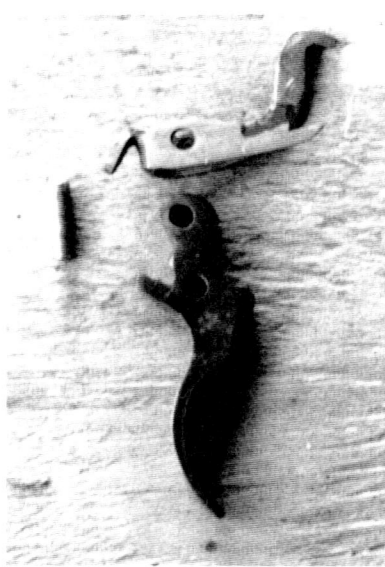

Trigger and trigger bar.

Trigger block.

Trigger with trigger bar pinned in place.

8 Grip

A piece of hardwood 1 1/2 x 5 x 2 1/2 inches is needed to make the pistol grip. It should preferably be a dense-grained wood such as maple, myrtle, walnut, or similar. Cabinet shops and furniture factories will very likely have a piece of scrap wood suitable for this. Lumberyards will surely have it, but you will very likely be compelled to buy an 8-foot board just to obtain a small piece. Stock blank suppliers also offer grip and forend blanks. While it is possible to make up such a grip using fiberglass or cast epoxy, this is an expensive, time-consuming process when only one unit is needed.

Assuming that a wood blank is used, a wide slot offering a snug fit over the frame is cut as deep as the frame height. The easiest way to do this is by using an end mill in the milling machine. The radiused corners can be formed with a small-diameter ball cutter. Although more labor is required, the same result can be achieved by making parallel saw cuts, as close together as possible, to the specified depth. Whatever excess material is left can be removed with a 1/2 inch wood chisel and finished with a file or rasp.

With the grip blank fitted in place over the frame, it is pushed forward against the end of the trigger guard strap and its location marked on the front side of the grip blank. A slot to fit is

cut using a 3/32 inch end mill. With the grip blank again in place and pushed forward over the end of the trigger guard strap, the bolt hole location is marked from the top through the existing bolt hole in the frame. This hole is drilled completely through the blank lengthwise using a 1/4 inch drill. After this, it is enlarged from the bottom side using a 9/16 inch drill to a depth of 1 7/8 inch. This provides clearance for the bolt head. The bottom shoulder of this counterbore should be flat. This is accomplished by grinding the drill point flat after the hole is almost to depth and finishing it with this. A 1/4 x 3 inch x 28 TPI bolt is adapted for use here by cutting a screw slot across the screw head, either by sawing or grinding with a small Dremel cutoff wheel.

The shape of the grip as shown feels comfortable, to me at least. I thought it had sort of a racy look. It can be reshaped to any individual taste. Forming to the desired shape can be accomplished most of the way with a disc sander. It is finished with a rasp, file, and sandpaper.

Sanding is accomplished using progressively finer grits of sandpaper. When you are satisfied with this, finish can be applied. This can consist of any number of finishes—oil, varnish, plastic, shellac, or whatever you like best. A durable, attractive finish, as used on the prototype gun, consisted of some 15 coats of Tru-Oil, a prepared oil finish sold in gun shops and sporting goods stores. Three or more coats are applied, drying between coats, and sanded back to the wood. This is repeated until all pores are filled. It is then given three or more coats and waxed.

If a "military" look is desired, several coats of flat black paint, filled and sanded as described, will give a finish closely resembling black plastic but will be far more durable.

Grip blank and finished grip.

Slot to accept lower receiver is cut in grip blank using milling machine. This can also be formed with a saw and chisel.

Finished grip mounted on frame.

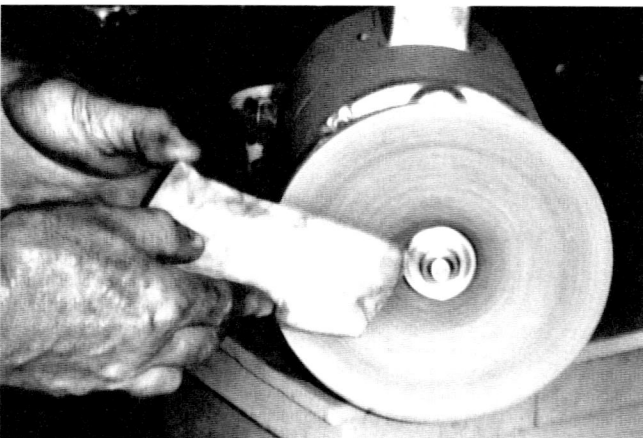

Grip is rough-shaped with disk sander and finished with rasps and sandpaper.

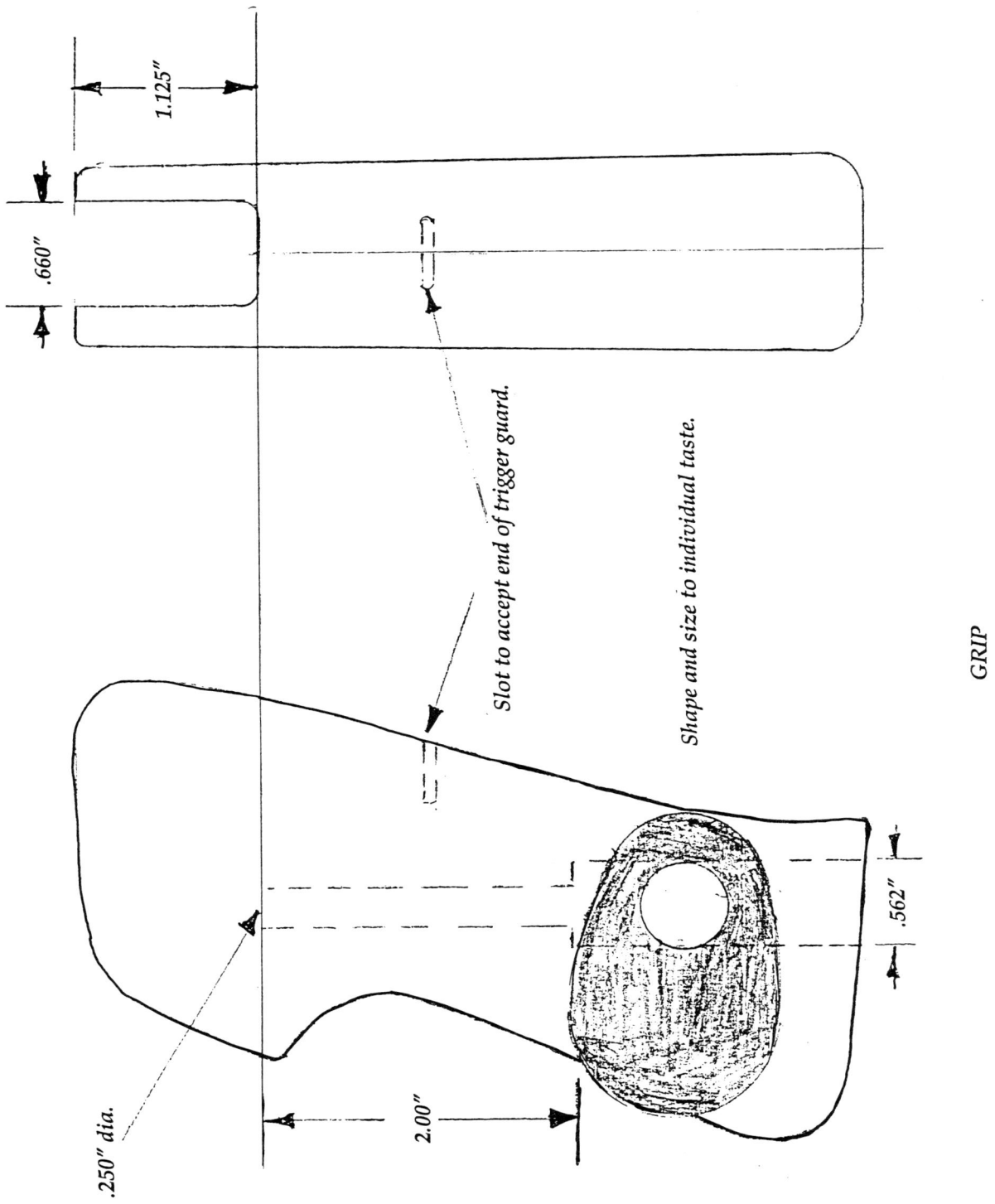

1.125"

.660"

Slot to accept end of trigger guard.

Shape and size to individual taste.

GRIP

.250" dia.

2.00"

.562"

9 Sights

While it is possible, perhaps desirable in some cases, to equip this gun with elaborate adjustable sights, in most instances the fixed sheet metal sights described here will suffice. The gun described herein has only a simple blade front sight and a fixed, notched rear. When these are aligned and cut to the proper height, they are completely adequate for their intended use.

The sights could be machined from solid blocks if desired, but it is far easier to form them from sheet metal blanks using a vise, a heavy hammer, and a pair of simple form blocks. The front blade and rear crosspiece are cut separately and silver soldered in place.

Form blocks, which can be used to form both the front and rear bases, are made up to form the protective ears and bases simultaneously. One block, which forms the curve or radius on the bottom side, is made from a short length of round stock of the same diameter as the receiver and approximately 2 inches long.

The other block should be curved on the bottom side to match the diameter of the other plus the thickness of the sheet metal used. In this case, it is 1 inch plus the thickness of the 16 gauge sheet metal, which is .0598 inch, or, for all practical purposes, .060 inch. This block should be .625 inch wide, .750 inch thick, and match closely the length of the other block.

Two 3/16 inch holes should be drilled, located near each end of the flat-sided form block from the top and extending all the way through. With both blocks clamped together, these holes are extended into the round blocks. Guide pins are positioned in the holes to keep the assembly in line while forming the parts.

In practice, a sheet metal blank is positioned between the two blocks and the assembly squeezed together using the vise. This will form the bottom contour. One upright side can be formed by forging with a block and hammer. The assembly is then turned over and again clamped in the vise, whereupon the other upright side is formed. The process is repeated to form the other base.

A lengthwise slot is cut down the center of one base to accept the front sight blade. The other will have a slot cut crosswise for the rear sight crosspiece. While it is easier to cut these in a milling machine using a 1/8 inch end mill, it can be done with a Dremel tool using a cutoff wheel and forming the ends with a small file.

Both blade and crosspiece should be snug fits into the slots. Ten gauge sheet metal is, supposedly, .1345, or .135, inch thick. This can be dressed slightly for a precise fit. Eleven gauge is approximately .120 inch in thickness. The slightly loose fit when using this can be eliminated by upsetting the edges of the slots slightly with a small punch.

The blades are fluxed and silver soldered in place. It may be necessary to recontour the bottom side of the bases slightly to remove excess silver solder or any protruding portion of the blades.

With flux applied to the adjoining surfaces, both sights are clamped in place and secured using silver solder. A piece of 5/8 inch wide material, long enough to fit across both sight bases, can aid in keeping both sights in line. By inserting it into the top sides of the sights and clamping it to the receiver, this will hold both sights in alignment while mounting. When joining steel gun parts, use silver solder with a 40 to 50 percent silver content. This will flow freely at a lower temperature than the low silver content crud used by plumbers, which is virtually useless for our purpose.

When cool, any residue generated as a result of the joining process should be removed and the rear sight notch cut. This will be regulated later, after assembly.

If you absolutely must have an adjustable rear sight, directions are given in my book *Home Workshop Guns for Defense and Resistance: Volume I, The Submachine Gun*. The Williams' Guide rear sight or its clone, the Marble #20, which is a steel copy of the aluminum Williams, can be mounted inside the rear base, assuming the crosspiece is removed, of course.

A simple form die is made by drilling a short piece of round stock of the same diameter as the receiver to accept two locating pins. A block of the desired width is radiused on the bottom side and holes are drilled to match the guide pins. Matching holes are drilled in the sheet metal blanks.

A blank is placed over the locating pins and the form block put in place.

Assembly is clamped in vise.

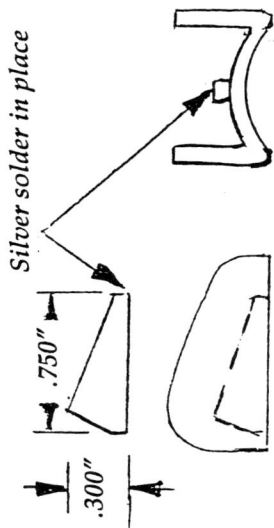

Silver solder in place

.750"

.300"

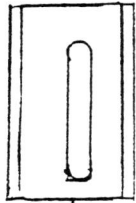

Slot .125" x .750"

1.625"

FRONT SIGHT

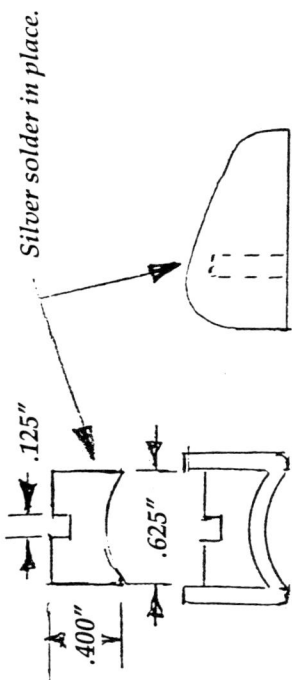

Silver solder in place.

.125"

.625"

.400"

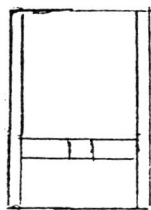

Slot .125" x .625"

.500"

.625"

1.100"

REAR SIGHT

Vise jaws are drawn together, forming curve on bottom.

Side is folded against block using block and hammer. Note that the assembly is supported underneath. This prevents it from moving downward in the vise due to hammer blows.

After one side is formed, the assembly is turned over and the other side is formed in the same manner.

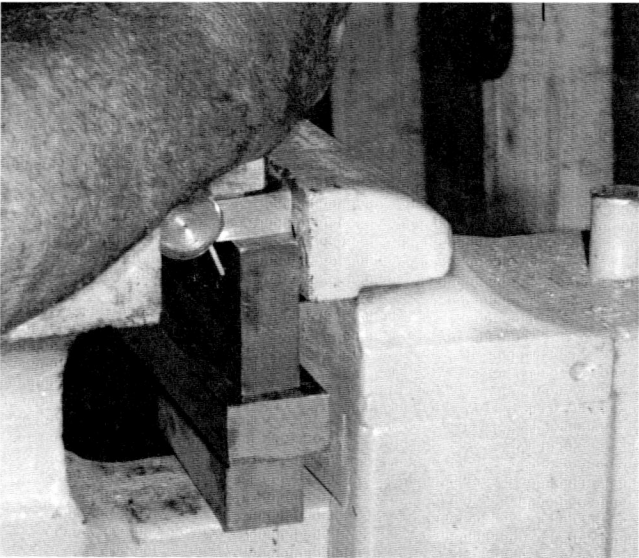

Rear sight base and insert.

Front sight base and blade.

Rear sight assembled.

Front sight assembled.

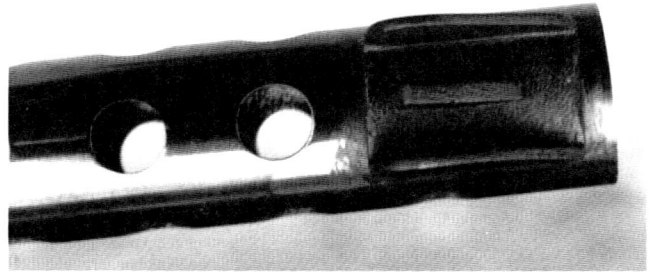

Front sight, top view.

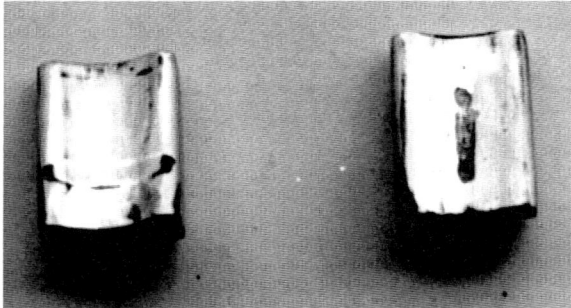

View from lower side showing blades in place.

Rear sight, top view.

Sights in place on receiver.

10 Assembly and Adjustment

With all the component parts completed, the gun should be assembled and tested before final polishing and bluing is done. Any final fitting and adjustment should be done at this time.

All working parts should have a smooth finish, with no burrs and scratches evident. Flat-sided parts, such as the sear and trigger, should have flat, smooth sides, square with the tops and bottoms, finished to a point where they feel slick when handled.

One way to achieve such a finish is by placing progressively finer sheets of abrasive cloth or paper on top of a piece of plate glass. (The glass provides a stiff, solid, smooth backing for the cloth.) The part to be polished is moved back and forth across the abrasive surface while maintaining downward pressure. An extremely fine, even finish can be obtained in this fashion.

When all of the interior parts are polished to your satisfaction, start assembly of the gun by installing both extractors, together with their springs, in the slots provided for them in the bolt and pinning them in place. If the closed-bolt version is used, the firing pin and striker are installed on the bolt. The barrel is inserted from the rear end and, with the indexing pin in alignment, pushed into place and secured by the barrel nut. The bolt is now inserted into the receiver and the cocking lever installed. With both the recoil spring and the striker spring in place, the recoil spring guide is installed, followed by the breech plug, which is held in place by the retaining pin.

The trigger group is assembled by first inserting the trigger block, with its return spring in place, into the lower receiver and pinning it in place. The trigger, with its trigger bar pinned in place, is put in place and pinned, making sure that the forward end of the trigger bar is under the projecting block at the top of the trigger block. If the two-stage trigger is used, the wire spring should be located and secured in place at this time. The trigger return spring is attached to the trigger bar, with its rear end stretched slightly to exert tension, and pinned in place at the rear of the lower receiver. The sear is now pinned in place as shown in the drawing. If only the closed-bolt version is used, the wire spring behind the trigger can be eliminated.

The upper receiver is positioned in place on the lower and secured in place using the front mounting screw. The grip is now positioned over the rear of the lower receiver and secured with the stock bolt, or grip screw, which also holds both receivers together at the rear. The magazine release, together with its spring, is pinned in place, completing assembly of the gun.

The closed-bolt gun can be tested for proper feeding with live ammunition, but the firing pin should be removed as a precaution against accidental firing. You should also determine that the chamber is cut deep enough to assure that the bolt face cannot crush the case rim as it slams forward and detonates the priming mixture. Don't neglect this; it can be dangerous. Once, I had a gun similar to this almost completed, except the firing pin had not been installed. I was going to show a friend of mine just how slick this gun would feed. I placed five cartridges in the magazine and inserted it in the gun. When I pulled the bolt to the rear and released it to feed the first cartridge, it fired all five rounds into the wall of my shop. Subsequent investigation revealed insufficient clearance between the bolt face and the cartridge head. This allowed the bolt to crush the priming mixture, firing the gun, without any firing pin whatsoever. Not only would the feds have claimed the gun was illegal, it was also dangerous and could have resulted in serious consequences.

The fixed firing pin in the open-bolt version prevents testing for feeding, but this can be done with the closed bolt in place.

With a single round in the magazine, draw the bolt to the rear and let it move forward smartly. If the cartridge feeds satisfactorily, try it with several rounds, working the action by hand. If cartridges do not feed properly, try to determine the cause by allowing the bolt to move forward slowly and observing where the bullet nose contacts the approach cone of the barrel. If it hits at the bottom, the forward ends of the magazine lips should be sprung open slightly. This will elevate the bullet nose in relation to the magazine body. If the bullet nose contacts the top of the barrel or the cartridge stands straight up, the magazine lips should be sprung inward, a little at a time, until the

condition is corrected. Be advised that when the bolt strips a cartridge from the magazine during normal firing, the bullet nose will try to move downward due to pressure being exerted against the upper rear of the cartridge case by forward movement of the bolt. So it probably won't take as much adjustment to the magazine lips as slow hand feeding may indicate.

When you are satisfied with the way the gun feeds during hand cycling, you are ready to test fire. If using the closed-bolt version, replace the firing pin assembly. One round of live ammunition should be loaded into the magazine. With the action cocked, hold the gun well away from your face and pull the trigger.

If everything works the way it should, the round will fire, causing the bolt to move to the rear, extracting and ejecting the empty case. The bolt should remain in the open position in the open-bolt version but should return forward in the closed-bolt.

If it worked the way it was supposed to, try it with two cartridges, still as a semiautomatic. We will get to the full automatic part soon, but some of the parts should be hardened first to prevent undue wear or battering.

If the bolt didn't remain open (open bolt), a little more fitting may be necessary. Try working the action by hand with the trigger depressed just far enough to release the bolt. The sear should catch and hold the bolt in its rearward position. If it does not, you may not have the trigger fitted or assembled correctly. Check it carefully.

If the trigger mechanism is working properly, then either the bolt is too heavy or the recoil spring is too stiff. In either instance, the breech block doesn't travel to the rear far enough for the sear to catch it. Try cutting one coil off the recoil spring and try it again, using one round as before. If it still doesn't remain open after firing, cut off another coil and try again. Repeat a third time if necessary.

If it still doesn't work after cutting off a third coil, something else must be wrong, or else you had an extremely stiff spring to begin with. Try polishing the bolt and the inside of the receiver body to reduce friction. If it still doesn't work properly, turn the bolt to a slightly smaller diameter (only .050 to .060 inch), leaving a full diameter band approximately 1/4 inch wide at each end and in the middle.

Take care not to weaken the spring or lighten the bolt so much that it will recoil far enough to the rear for the cocking lever to strike the end of its slot. To check against this happening, wrap a layer of tape around the receiver, covering the last 1/4 inch of the cocking lever slot, and fire the gun. If the tape isn't torn by the cocking lever all the way to the end of the slot, it can be considered satisfactory. If it does, a slightly stronger spring is needed.

When you are satisfied that you have it adjusted and working properly, try firing with two rounds in the magazine. The trigger must be released and pulled again to fire subsequent shots. Anything else is unacceptable and must be corrected.

Assuming that it does work correctly, the gun should now be disassembled and the parts heat treated, as described in the next chapter. After this is done, assemble the gun once more and test fire it thoroughly, first on semiautomatic fire, and then on full automatic.

When testing as a full automatic, start by loading only two or three rounds in the magazine. This will prevent having a runaway gun if something should break or fail to work properly. It isn't my idea of fun to have a full-automatic weapon with a full magazine continue to fire after the trigger is released. If it should, all you can do is hold onto it and hope it runs dry before you hit anybody. So test it thoroughly with only a few rounds in the magazine before stuffing it full.

11 Heat Treatment

I have already gone through the basic principles of heat treatment in at least two other books, so there isn't any reason to repeat it here. Suffice to say, several of the small parts described herein will require hardening. Please note that all temperatures given are in Fahrenheit. I have a hard enough time even spelling it, let alone writing it repeatedly. In certain instances, this hardening process is required only to prevent wear and in others both to increase strength and prevent battering or other malformation.

It will be necessary for you to heat the part to be hardened to a temperature above the critical stage (forming Austenite), then rapidly cooling it by plunging it into a quenching bath consisting of oil, water, brine, etc. (forming Martensite). The extremely hard steel is then heated once more to a temperature somewhere between 300 and 1290 degrees and cooled (forming either Troosite or Sorbite). The exact temperature required for this tempering or drawing operation varies considerably, depending both on the carbon content of the steel and the strength and hardness requirements.

A gas or electric furnace is almost a necessity for this type of heat treatment. If you anticipate treating many parts, I suggest you either buy a commercial furnace or build one. A usable gas furnace can be built by simply lining a steel or iron shell with firebrick. A vacuum cleaner motor and fan can be used as a blower. A pyrometer is also necessary to measure and regulate the temperature.

Don't depend on commercial heat treaters to treat your parts correctly. We have a loudmouth here in town who brags about his abilities and knowledge of heat treatment. Some time ago I let him talk me into having him heat treat 10 sets of shotgun parts. I asked that they be quenched at 1600 degrees and drawn at 800. When I went back for the parts, he went through an elaborate line of nonsense about how he had packed the openings with steel wool and taken all sorts of precautions to prevent distortion and how much better these would be than any I had used before. He also charged me about five times what the job was worth. I didn't complain; since I hadn't inquired as to the price, I felt it was my own fault. But when I test fired the guns, the bolts cracked. Luckily, I discovered this before they were all ruined. I managed to save three. But the ones that cracked simply shattered like glass when hit with a hammer. When I confronted the self-proclaimed expert with the ruined parts, he went into a lengthy spiel about how it wasn't his fault and finally refunded what he had charged me. I retempered the remaining parts and managed to save them. But it took several hours of machine work to replace the ruined ones. Since then, I have done my own work.

It is possible to harden and temper parts by using the flame of an oxy-acetylene torch, a forge, and, in some instances, a hot bath. The latter method can be either a chemical solution or molten metal. This method is especially well-suited to irregularly shaped parts, parts with holes, and parts varying in thickness or mass. These parts will heat uniformly to the desired temperature in such a bath.

There are times, however, when the only available method will be the torch. While this method is far from foolproof, satisfactory results can be obtained if sufficient care is taken.

In many cases you will not know the exact composition of your steel, so a bit of exper-imenting is in order before beginning. Since most of the medium- and high-carbon steels require heating to between 1400 and 1650 degrees for hardening, try heating the scrap to a bright, clear, glowing red, devoid of any yellow-ish tinge.

This is the "cherry red" so often mentioned in connection with heat-treating activities. When this color is reached, the material is plunged into a quenching bath of water or SAE #10 motor oil, which is at room temperature or slightly warmer. It should now be so hard, a file won't touch it. If it is not, try another scrap at a slightly higher temperature, and when the proper combination is found, apply it to the part to be hardened.

Nearly all carbon steels change color in the same way and at almost the same temperatures. So the hardening and tempering colors that appear while heating will indicate the approx-imate temperatures of the metal. The chart at the end of this chapter gives a fairly broad color range and can be used as a guide.

There is a product on the market called Tempilaq which can take some of the guesswork out of the temperature control. It is available from gunsmith supply houses and machine tool suppliers. In use, a thin coating is applied to the surface of the material to be heat treated. Actually, only a thin smear is required. After it

dries to a dull finish, begin heating the metal. The Tempilaq will melt sharply when the proper temperature is reached and should be quenched immediately. This product is available to indicate temperatures from 350 to 1550 degrees and, except for expensive pyrometers, is the most accurate temperature indicator I have found.

Whatever type of temperature indicator is used, the hardened steel must be tempered, or drawn, after quenching. Either wipe on a smear of Tempilaq or heat the metal to the color indicating the temperature desired, then allow it to cool. It may be wise to again experiment with a hardened scrap of the same material before attempting to temper the actual part, and test it again with a file and punch.

Another method which could prove useful for drawing temperatures up to 500 degrees is the use of the kitchen oven. Simply place the parts in the oven, set it to the desired temperature, and let them heat for 30 minutes to an hour.

Still another method that works well on firing pins, sears, pins, and other small parts is the use of a hardening compound such as Kasenit. By heating the part to be hardened to a cherry red and coating it with the compound, usually done by burying the part in the compound, then reheating to the same cherry red and quenching in water, a hard surface will result while retaining a soft inner core. This is similar to a case-hardening process, which I will not attempt to explain here since this process will give similar results with less equipment. The directions on the can should be followed exactly if this method is used, since different metals require different treatment.

It may be helpful to include a brief break-down of the SAE numbers used in drawings and specifications to indicate a certain kind of steel. We read about 2340, 4340, 1035, etc., which, to the average person, means little or nothing. The first figure, as a general rule, indicates the class to which the steel belongs. Thus, 1 indicates a carbon steel, 2 a nickel steel, 3 nickel chromium, 4 molybdenum steel, 5 chromium steel, 6 chrome vanadium steel, etc.

In the case of the alloy steels, the second figure generally indicates the approximate percentage of the alloying element. Usually, the last two or three figures indicate the average carbon content in hundredths of one percent, or "points." Thus, 2340 means a nickel steel with approximately 3 percent nickel and 0.40 (forty hundredths) percent carbon.

The following color chart may come in handy when tempering by the color method. Brightly polish the part to be tempered so that the color will show and place it on a red-hot steel plate until it reaches the desired color, then remove it and allow it to cool.

It should be remembered that the methods and descriptions in this chapter apply to carbon steels only. Certain alloy steels may require

Surface hardening compounds such as Kasenit can be used to impart a hard surface to parts made of low carbon steel while retaining a soft core. Useful mainly on small parts.

entirely different methods of heat treatment. Also remember that since I have no control over your attempts at heat treatment, I cannot accept any responsibility for problems you may encounter.

Hardening & Tempering Colors	Degrees F	Tempilaq Available
Pale Yellow	425	400-413-425
Pale Straw	450	438-450
Yellowish Brown	500	475-488-500
Light Purple	525	525
Purple	530	
Blue	550	550
Dark Blue	600	575-600
Bluish Green	625	650
Barely Visible Red	900	
Blood Red	1200	
Cherry Red	1400	1350-1400-1425
Light Red	1500	1480-1500
Orange	1650	
Yellow	1800	
Light Yellow	2000	
White	2200	

12 Finishing and Bluing

All visible metal surfaces should be polished to a point where they are smooth and free from tool marks or other blemishes. This can be accomplished using power buffing wheels, which is considerably faster, or by hand polishing with files and abrasive cloth.

Files should be used to remove tool marks, dents, etc. Fairly coarse abrasive cloth is then applied, using strips of the cloth in a "shoeshine" fashion around the curved surfaces of the barrel, bolt, and receiver alike. Dents and low places will be revealed as this progresses.

This cross polishing is followed by lengthwise polishing whereby strips of the abrasive cloth are wrapped around files or blocks and moved lengthwise along the metal, parallel to the longitudinal axis, turning them slightly as the polishing progresses. This is continued until the entire area has been gone over. Lateral depressions and circular tool marks will become apparent as this is done.

This process is continued crosswise followed by lengthwise polishing until all dents, pits, depressions, and tool marks are removed. Finally, after carefully polishing with the finest grit available, polish all surfaces with the crocus cloth. Use the cloth in both directions, but finish with lengthwise strokes as you did with the coarser grades as described previously.

Power polishing is done in the same manner. Begin by applying a coarse-grit abrasive compound to either felt or cloth wheels. This is followed by progressively finer grits until the desired degree of finish is reached. Felt wheels should be used when polishing over screw or pin holes and on flat surfaces, especially where straight lines and sharp corners must be maintained. When using the power wheels, crosswise polishing should be avoided whenever possible. The parts should be held at an angle to the wheel and polished lengthwise whenever possible.

When polished to your satisfaction, the parts should be examined in direct sunlight to ascertain that no scratches or polishing marks remain. Following this final check, the individual parts should be degreased. While at least 50 percent of obtaining a good blue job depends on the quality of the polish, another 25 percent will depend on the parts being absolutely free of any trace of oil or grease.

A number of degreasing compounds and detergents are available in grocery, paint, and hardware stores. Mix one of these with water and boil the parts in the solution for a few minutes. After rinsing in clear water, they will be ready for the bluing process. After this, the parts should no longer be handled with bare hands since the oil in the skin of your hands may contaminate them. From this point, use rubber gloves, metal hooks, or wires to handle them.

In previous volumes I have given detailed descriptions, including formulas, for both nitrate bluing, or "blacking," and rust blue methods. In this book I will try to describe a method sometimes referred to as "fume bluing" or "fuming." This method is probably the simplest and most foolproof and requires less equipment than most.

Along with a tank to boil the parts in plus a suitable heat source, it is necessary to have at least one, preferably two, plastic boxes, both as airtight as possible, to contain the parts while the actual fuming takes place. One box must be of sufficient size to accept the barreled receiver; the other needs only to be of adequate size to hold the remaining parts. Sign shops often have scraps of plastic sheet or discarded signs made from 1/8 to 1/4 inch thick plastic from which a top, bottom, sides, and ends can be sawn and, using suitable cement, assembled into satisfactory receptacles.

You will also need a small quantity of both concentrated nitric and hydrochloric acids, as well as several (six to eight) plastic cups to hold these acids.

With the parts degreased and rinsed, rubber plugs or corks are placed in each end of the barrel. Any areas not to be blued can be masked off or coated with shellac, varnish, lacquer, etc. The parts are now placed inside the plastic boxes. Several drops of each acid are placed in separate cups (don't mix them) and two cups of each acid placed in the long box along with one or two of each in the smaller box and the covers put in place. The actual coloring usually takes place in one to three hours. Therefore, the work should be examined frequently after the first hour and removed when the desired color is obtained. Making the box lids from clear plastic can be an aid to easy inspection.

When finished, the parts are boiled in clean water to stop any further action and oiled in the same manner used with other methods.

It is possible to obtain about any degree of luster desired by varying the acid quantities, since the nitric acid does the actual bluing while the hydrochloric fumes simply etch the surface. A bit of experimenting is necessary to achieve the desired finish.

This method was used to color all parts of the prototype gun pictured herein except the lower receiver. This was given an imitation "case-hardened" color which contrasts with the blued parts, breaking up what might be considered by some as an overabundance of blued metal. This, coupled with the grip of figured English walnut, resulted in an eye-catching gun that was considered most attractive by most of those who have seen it.

This finish, along with a number of others, is detailed in my book, *Home Workshop Prototype Firearms*. It consists simply of painting a coat of tincture of benzoin on the polished steel and allowing it to dry. Small spots or strips are then heated quickly with an oxy/acetylene torch until the color appears and quenched in water. Another adjacent area is treated in the same manner and repeated until the entire surface is treated. The material must not be allowed to overheat or the color will char and flake off. Upon completion, a coat of varnish or other sealer should be applied to protect the colors.

Hot nitrate bluing method can be used if available. For the method is described, only a hot water tank is required.

60

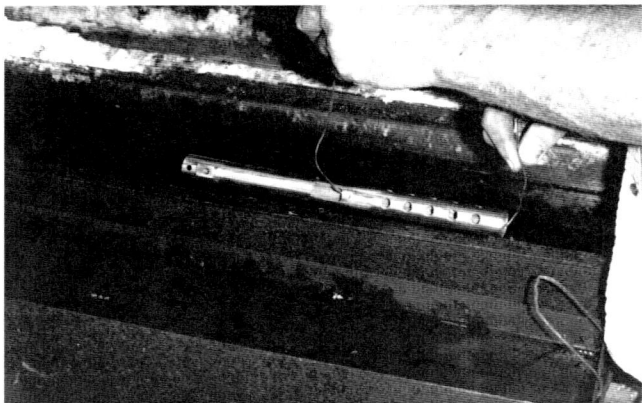

After degreasing, parts should not be handled with bare hands.

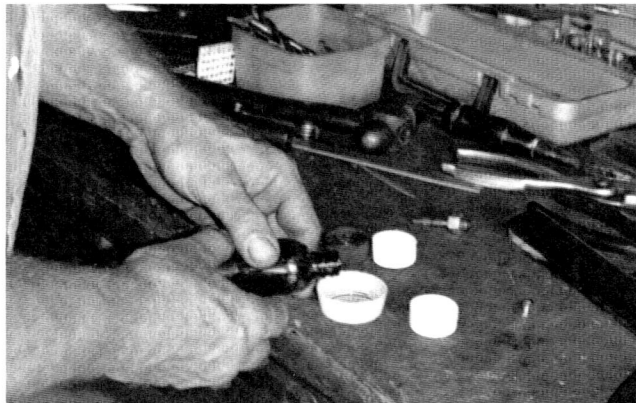

Acid is poured into small plastic cups. Only a small amount is required.

A hard felt wheel should be used for primary power polishing to avoid dishing holes and rounding sharp corners and edges.

Plastic cake boxes, available at stores such as Wal Mart, can be used for pistol-sized projects. Note that four acid containers are used here, two for each acid.

Final polishing is done with sewed cotton wheels.

With cover in place, fuming action should be allowed to work for an hour or more. Parts should then be examined frequently and removed when satisfactory color is obtained.

The finished product, assembled and ready for use.

*P*ostscript

After the prototype gun was completed, we shot it several hundred times and discovered that its accuracy was above average. I then decided that a better trigger pull could be achieved through use of a hammer-fired ignition system. With this in mind, I built another gun which worked out well enough that we decided it should be included as an option in this book.

The pin hole locations in the lower receiver must be moved to accommodate the altered trigger parts. Actually, this may be an advantage, since both the hammer and sear pins are now covered by the grip, leaving only the trigger pin exposed. Note that the hammer pin is 5/32 inch, or .156 inch, in diameter, while the sear and trigger pins remain .125 inch, as before.

The bolt uses the same general dimensions as the open bolt except that the recoil spring pocket is moved to the left side (when viewed from the rear) as far as possible and an opening made to accept the moving firing pin. The firing pin hole must be drilled off center enough so that the firing pin will strike the outer rim of the cartridge case. This can be accomplished in the lathe by using a four-jaw chuck and offsetting the bolt the proper amount. However, since this hole must be started with a 1/8 inch end mill, due to the fact that a drill would be un-supported on one side at the bolt face end, it will be easier to exchange the end mill with a 3.2mm drill and continue drilling the hole with the milling machine. The fore mentioned drill is .126 inch in diameter, which allows a small amount of clearance for the firing pin. The rear end of the bolt is machined to provide clearance for the enlarged rear end of the firing pin as well as the return spring.

The firing pin can be made from 1/8 inch drill rod with the enlarged end piece silver soldered in place. Both sides of the forward end should be dressed to a flat wedge shape, leaving the tip .040 to .050 inch wide. The firing pin should be cut slightly longer than required and hand finished so that when the rear end is flush with the base of the bolt, the tip protrudes past the bolt face some .040 inch.

A cross pin hole is drilled through the bolt body with the firing pin in place. This will assure alignment. The resulting hole through the firing pin base is lengthened to form a slot .200 inch long. Properly done, this will allow the firing pin to move fully forward with the retaining pin in place and to retract some .075 inch when no forward pressure is exerted against it. The firing pin, when pinned in place, should move freely, without binding.

The trigger is cut to shape from 1/4 inch flat stock. Except for the dimensions specified in the drawing, the remainder can be changed to whatever shape and size desired.

The trigger bar is cut from 14 gauge sheet stock. The pin hole should be located and drilled after the part is formed to shape. Both the hook that engages the sear and the disconnector portion should be left oversize and fitted during assembly. This part is surface hardened when finished using Kasenit or something similar.

The hammer and sear should be made from a tough, shock-resistant steel such as 4140. The corresponding hammer and sear notches should be left oversize and hand fitted during assembly. Suitable hammer springs intended for use in cheap shotguns and rifles can sometimes be located. If none are available commercially, a "mousetrap" type spring can be formed using music wire of .045 to .050 inch in diameter. It is also possible to straighten the two outside coils on each side of an M16/AR-15 hammer spring and shorten the resulting legs. This will reduce the width of the spring, allowing it to function inside the narrow lower receiver

The safety is made in two parts, three actually, if you count the connecting piece which is silver soldered to the upper part. A slot is cut through the bottom of the lower receiver to allow the safety to slide back and forth. The parts are pinned together as shown. The detent at the bottom of the trigger return spring engages in dimples at each end of the safety's movement, providing friction, or resistance, to hold it in place.

The above will result in a trigger with a light pull and very little travel, allowing the full accuracy potential of the gun to be utilized.

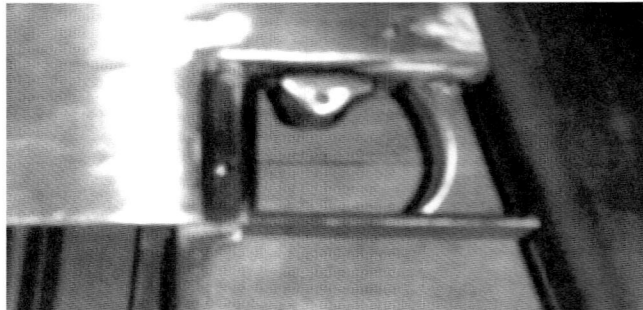

This photo, taken before final polishing, shows safety in position just forward of the trigger. Pushing the safety toward the trigger causes it to engage the sear, preventing firing. When pushed forward, the sear is released.

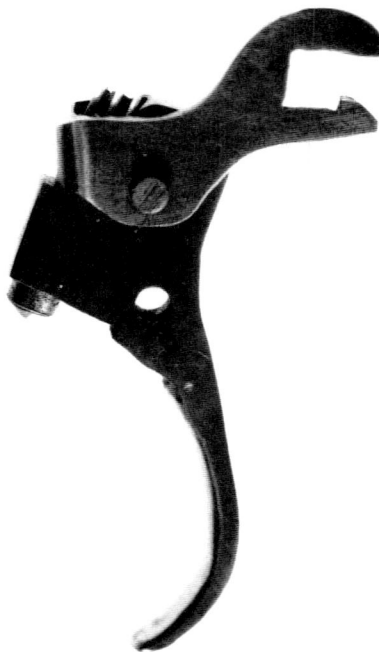

Trigger with trigger bar in place.

The safety is made in two parts and pinned together.

Sear shown with spring in place.

Hammer with spring in place.

The bolt must have an offset recoil spring. Full-length firing pin.

A hammer spring can be formed from spring wire, made from existing springs such as those used in various cheap shotguns and rifles, or cut down from an M16 spring or something similar.

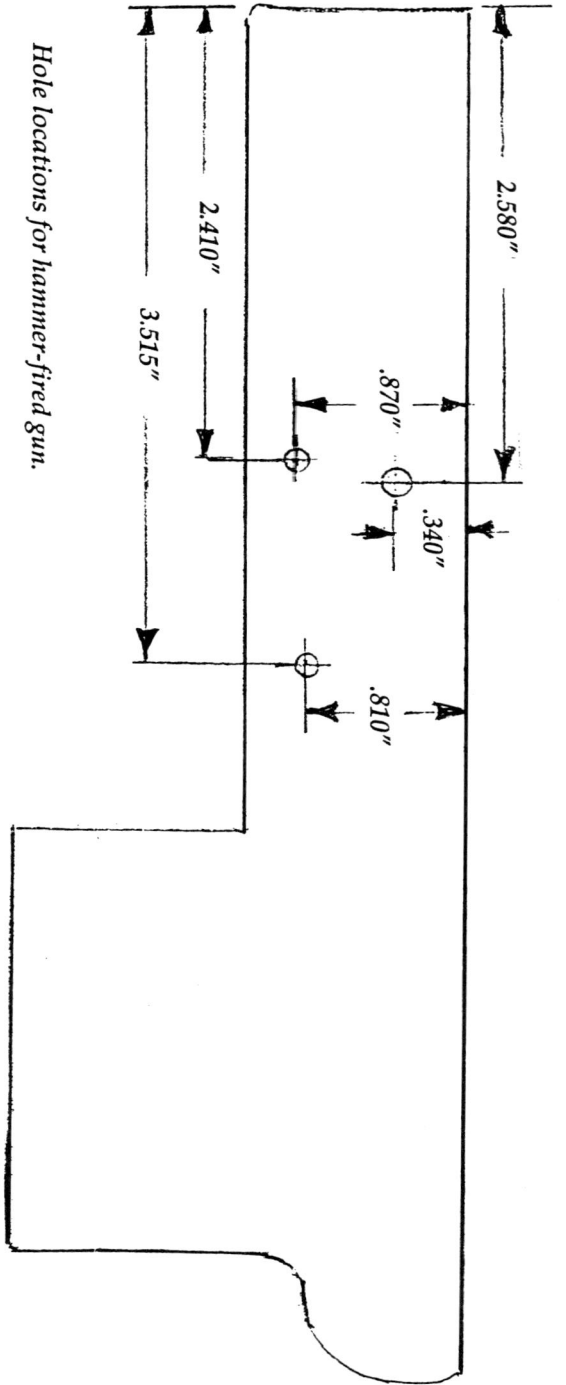

Hole locations for hammer-fired gun.

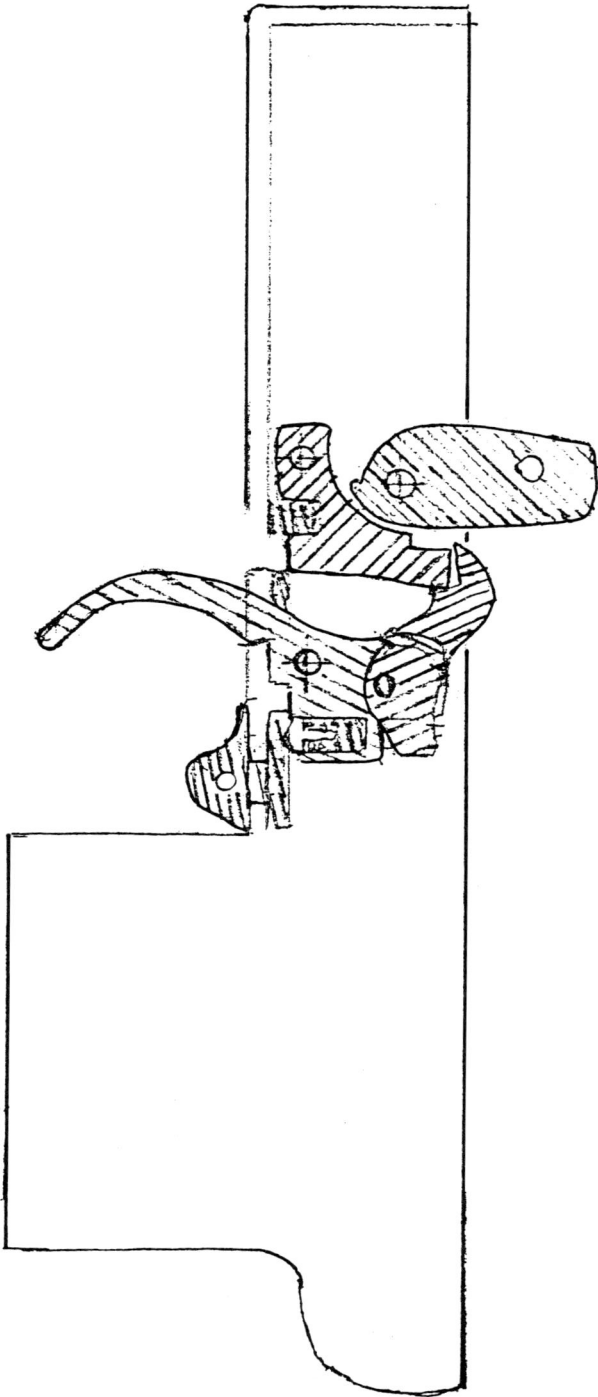

2.580"

2.410"

3.515"

.870"

.340"

.810"

HAMMER-FIRED TRIGGER ASSEMBLY.

.375"

.375"

.750"

.093" hole, drilled through both parts simultaneously after assembly

Dimples formed with .125" drill to mate with detent on end of trigger spring

.315"

.125"

.425"

Slot .125" wide x .315" long

.125"

.125"

.450"

1.375"

Slot .270" x .500" to provide trigger clearance

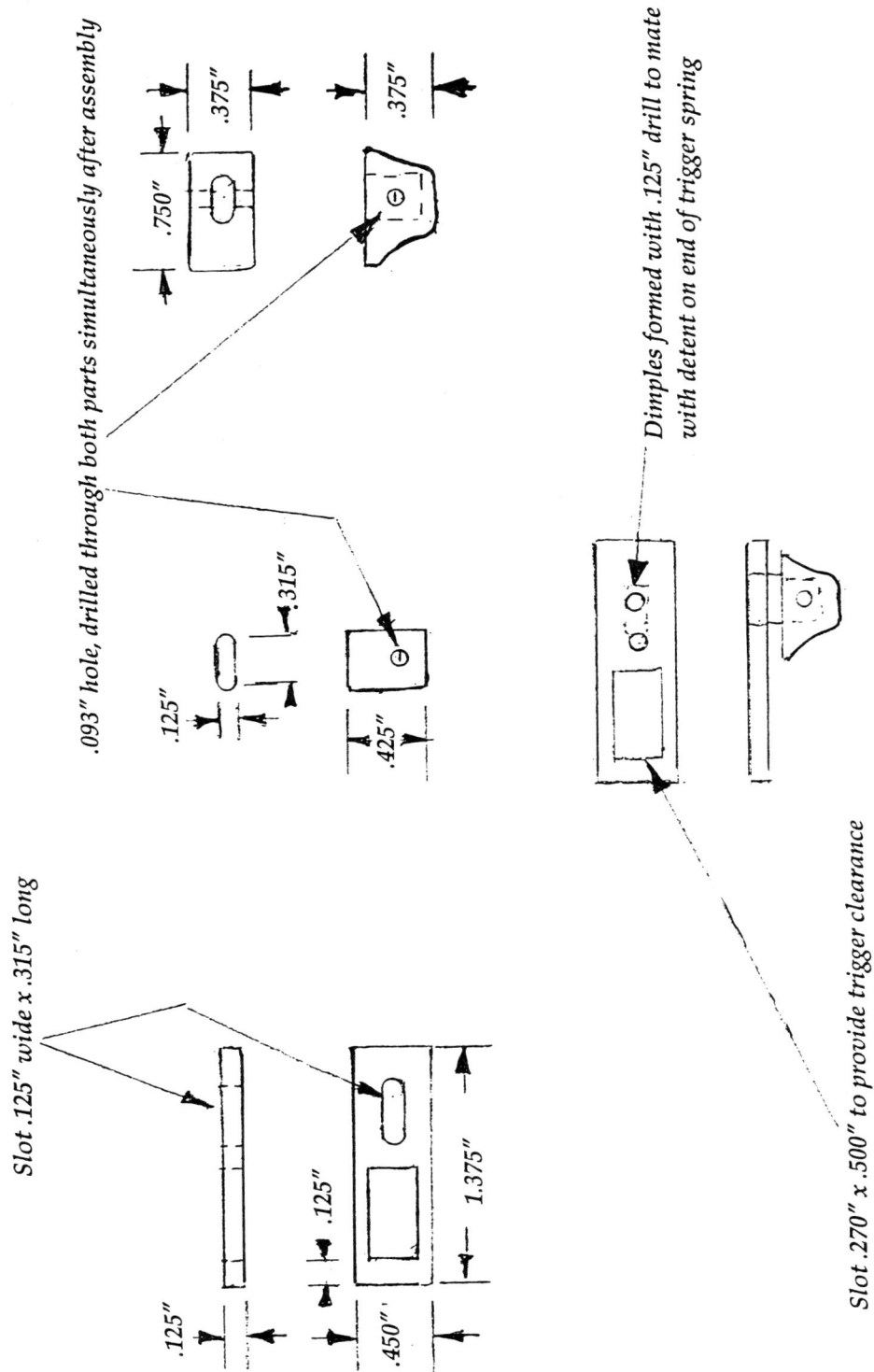

SAFETY ASSEMBLY

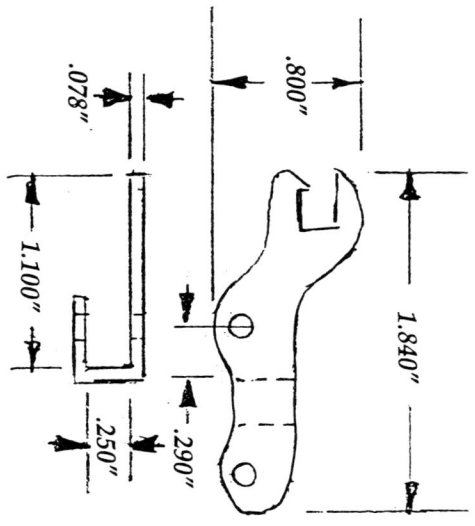

TRIGGER BAR

.078"

.800"

1.100"

.250"

.290"

1.840"

TRIGGER

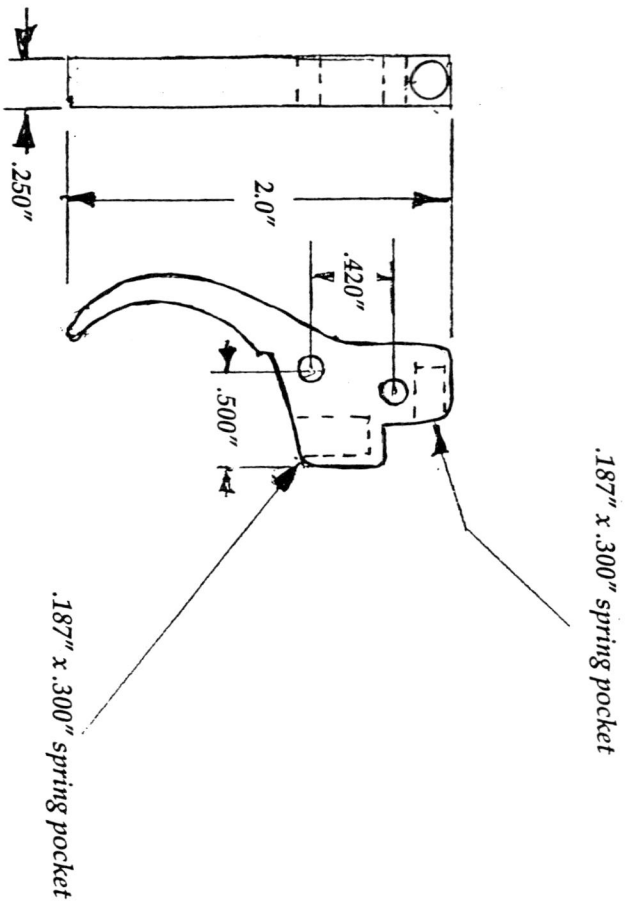

.250"

2.0"

.420"

.500"

.187" x .300" spring pocket

.187" x .300" spring pocket

BUSHING

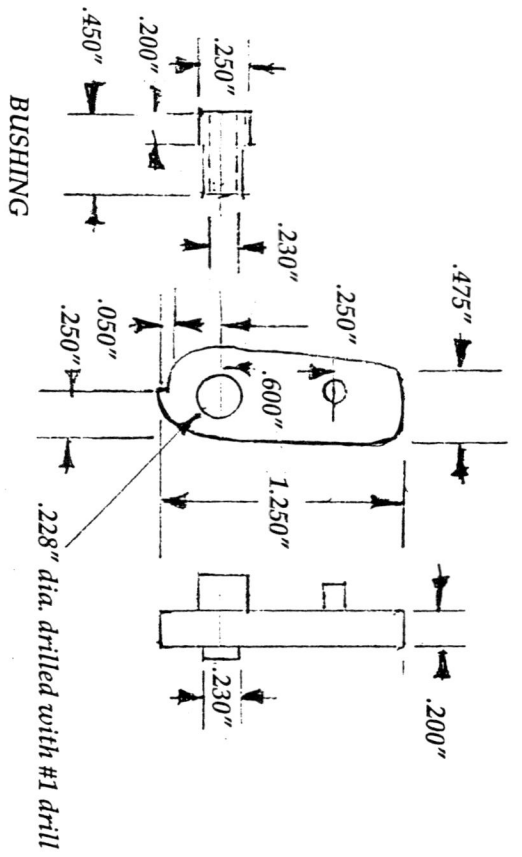

.490"

.100"

.900"

.450"

.200"

.250"

.230"

.050"

.250"

.475"

.250"

.600"

1.250"

.230"

.200"

.228" dia. drilled with #1 drill

HAMMER

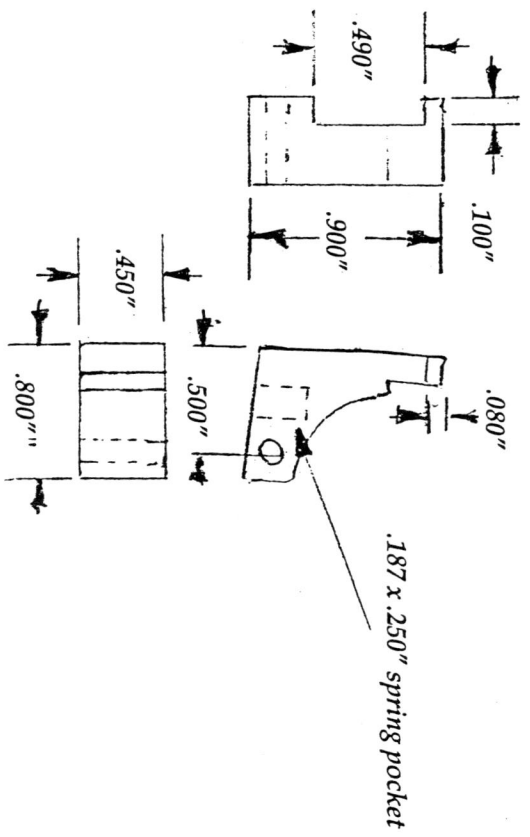

.800""

.500"

.080"

.187 x .250" spring pocket

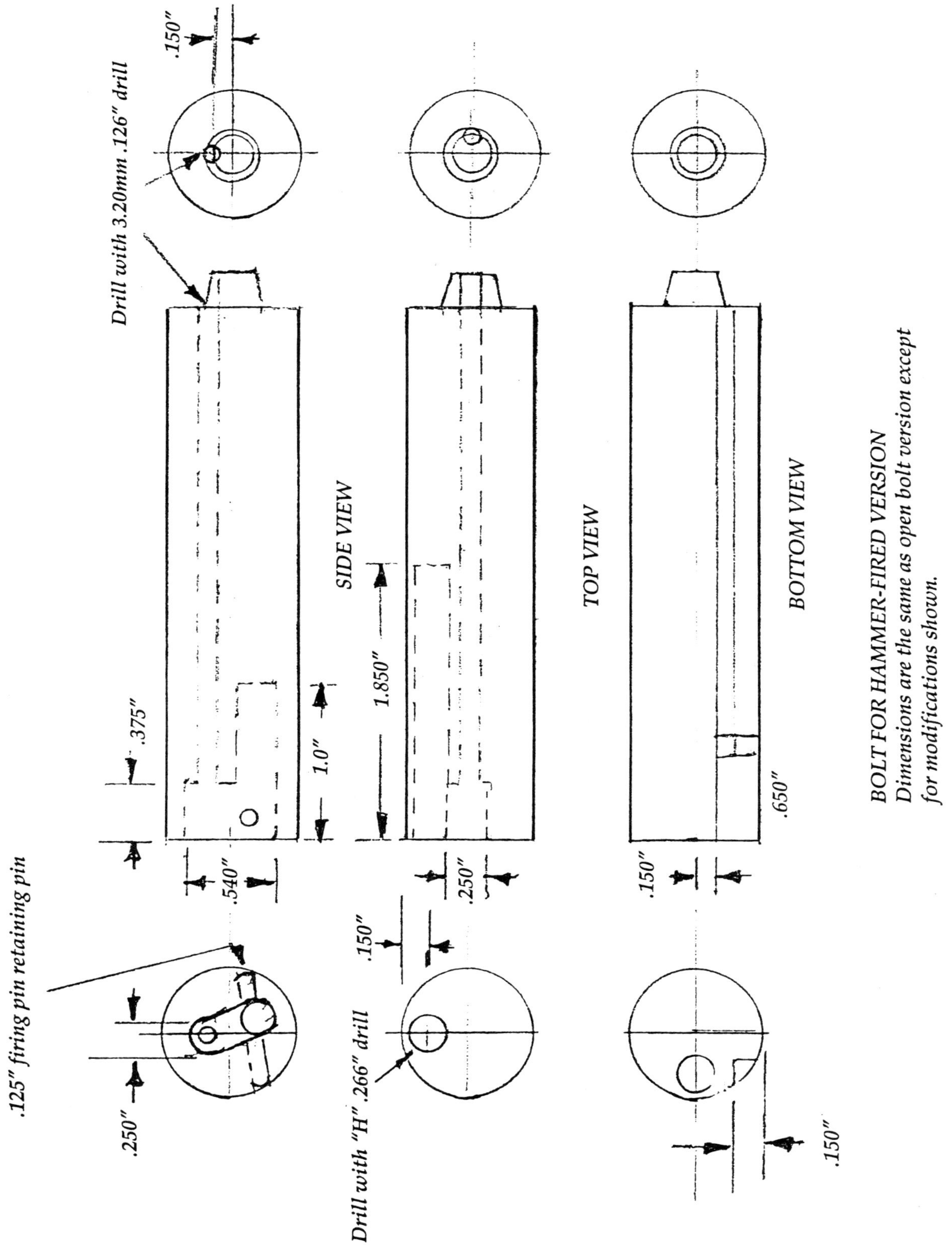

Drill with 3.20mm .126" drill

.150"

SIDE VIEW

TOP VIEW

BOTTOM VIEW

BOLT FOR HAMMER-FIRED VERSION
Dimensions are the same as open bolt version except for modifications shown.

.375"

1.850"

1.0"

.540"

.250"

.150"

.650"

.125" firing pin retaining pin

.250"

.150"

Drill with "H" .266" drill

.150"

.150"

Firing pin retainer slot .125" x .200"

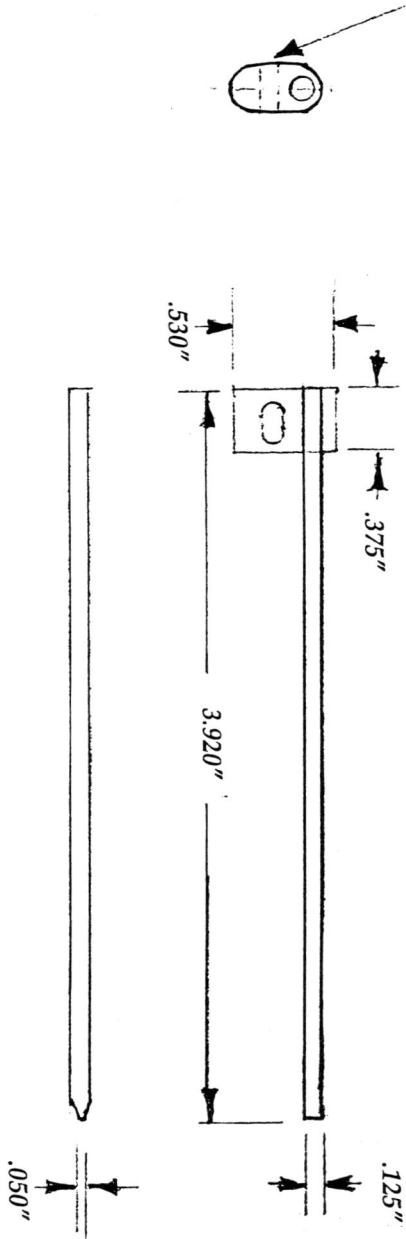

.150"

.125" dia.

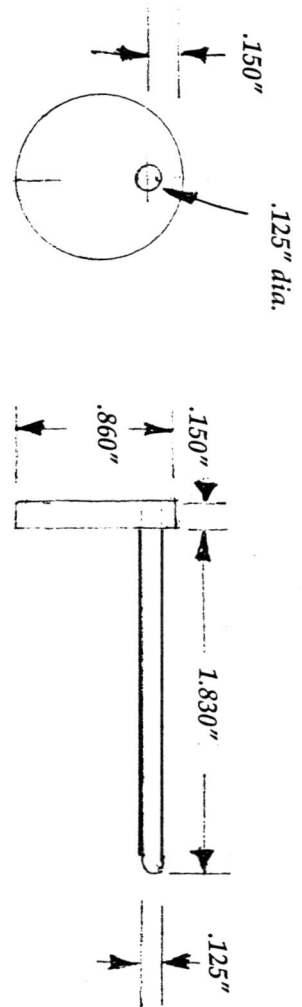

RECOIL SPRING GUIDE

.860"

.150"

1.830"

.125"

FIRING PIN

.530"

.375"

3.920"

.125"

.050"